IN ASSOCIATION WITH SQA

HODDER GIBSON

Model Papers WITH ANSWERS

PLUS: Official SQA 2014 & 2015 Past Papers With Answers

National 5 Physics

Model Papers, 2014 & 2015 Exams

This book contains the official SQA 2014 and 2015 Exams for National 5 Physics, with associated SQA approved answers modified from the official marking instructions that accompany the paper.

In addition the book contains model papers, together with answers, plus study skills advice. These papers, some of which may include a limited number of previously published SQA questions, have been specially commissioned by Hodder Gibson, and have been written by experienced senior teachers and examiners in line with the new National 5 syllabus and assessment outlines, Spring 2013. This is not SQA material but has been devised to provide further practice for National 5 examinations in 2014 and beyond.

Hodder Gibson is grateful to the copyright holders, as credited on the final page of the Answer Section, for permission to use their material. Every effort has been made to trace the copyright holders and to obtain their permission for the use of copyright material. Hodder Gibson will be happy to receive information allowing us to rectify any error or omission in future editions.

Hachette UK's policy is to use papers that are natural, renewable and recyclable products and made from wood grown in sustainable forests. The logging and manufacturing processes are expected to conform to the environmental regulations of the country of origin.

Orders: please contact Bookpoint Ltd, 130 Park Drive, Milton Park, Abingdon, Oxon OX14 4SE. Telephone: (44) 01235 827720. Fax: (44) 01235 400454. Lines are open 9.00–5.00, Monday to Saturday, with a 24-hour message answering service. Visit our website at www.hoddereducation.co.uk. Hodder Gibson can be contacted direct on: Tel: 0141 848 1609; Fax: 0141 889 6315; email: hoddergibson@hodder.co.uk

This collection first published in 2015 by
Hodder Gibson, an imprint of Hodder Education,
An Hachette UK Company
2a Christie Street
Paisley PA1 1NB

Typeset by Aptara, Inc.

Printed in the UK

A catalogue record for this title is available from the British Library

ISBN: 978-1-4718-6066-9

3 2 1

2016 2015

Introduction

Study Skills – what you need to know to pass exams!

Pause for thought

Many students might skip quickly through a page like this. After all, we all know how to revise. Do you really though?

Think about this:

"IF YOU ALWAYS DO WHAT YOU ALWAYS DO, YOU WILL ALWAYS GET WHAT YOU HAVE ALWAYS GOT."

Do you like the grades you get? Do you want to do better? If you get full marks in your assessment, then that's great! Change nothing! This section is just to help you get that little bit better than you already are.

There are two main parts to the advice on offer here. The first part highlights fairly obvious things but which are also very important. The second part makes suggestions about revision that you might not have thought about but which WILL help you.

Part 1

DOH! It's so obvious but …

Start revising in good time

Don't leave it until the last minute – this will make you panic.

Make a revision timetable that sets out work time AND play time.

Sleep and eat!

Obvious really, and very helpful. Avoid arguments or stressful things too – even games that wind you up. You need to be fit, awake and focused!

Know your place!

Make sure you know exactly **WHEN and WHERE** your exams are.

Know your enemy!

Make sure you know what to expect in the exam.

How is the paper structured?

How much time is there for each question?

What types of question are involved?

Which topics seem to come up time and time again?

Which topics are your strongest and which are your weakest?

Are all topics compulsory or are there choices?

Learn by DOING!

There is no substitute for past papers and practice papers – they are simply essential! Tackling this collection of papers and answers is exactly the right thing to be doing as your exams approach.

Part 2

People learn in different ways. Some like low light, some bright. Some like early morning, some like evening / night. Some prefer warm, some prefer cold. But everyone uses their BRAIN and the brain works when it is active. Passive learning – sitting gazing at notes – is the most INEFFICIENT way to learn anything. Below you will find tips and ideas for making your revision more effective and maybe even more enjoyable. What follows gets your brain active, and active learning works!

Activity 1 – Stop and review

Step 1

When you have done no more than 5 minutes of revision reading STOP!

Step 2

Write a heading in your own words which sums up the topic you have been revising.

Step 3

Write a summary of what you have revised in no more than two sentences. Don't fool yourself by saying, "I know it, but I cannot put it into words". That just means you don't know it well enough. If you cannot write your summary, revise that section again, knowing that you must write a summary at the end of it. Many of you will have notebooks full of blue/black ink writing. Many of the pages will not be especially attractive or memorable so try to liven them up a bit with colour as you are reviewing and rewriting. **This is a great memory aid, and memory is the most important thing.**

Activity 2 – Use technology!

Why should everything be written down? Have you thought about "mental" maps, diagrams, cartoons and colour to help you learn? And rather than write down notes, why not record your revision material?

What about having a text message revision session with friends? Keep in touch with them to find out how and what they are revising and share ideas and questions.

Why not make a video diary where you tell the camera what you are doing, what you think you have learned and what you still have to do? No one has to see or hear it, but the process of having to organise your thoughts in a formal way to explain something is a very important learning practice.

Be sure to make use of electronic files. You could begin to summarise your class notes. Your typing might be slow, but it will get faster and the typed notes will be easier to read than the scribbles in your class notes. Try to add different fonts and colours to make your work stand out. You can easily Google relevant pictures, cartoons and diagrams which you can copy and paste to make your work more attractive and **MEMORABLE**.

Activity 3 – This is it. Do this and you will know lots!

Step 1

In this task you must be very honest with yourself! Find the SQA syllabus for your subject (www.sqa.org.uk). Look at how it is broken down into main topics called MANDATORY knowledge. That means stuff you MUST know.

Step 2

BEFORE you do ANY revision on this topic, write a list of everything that you already know about the subject. It might be quite a long list but you only need to write it once. It shows you all the information that is already in your long-term memory so you know what parts you do not need to revise!

Step 3

Pick a chapter or section from your book or revision notes. Choose a fairly large section or a whole chapter to get the most out of this activity.

With a buddy, use Skype, Facetime, Twitter or any other communication you have, to play the game "If this is the answer, what is the question?". For example, if you are revising Geography and the answer you provide is "meander", your buddy would have to make up a question like "What is the word that describes a feature of a river where it flows slowly and bends often from side to side?".

Make up 10 "answers" based on the content of the chapter or section you are using. Give this to your buddy to solve while you solve theirs.

Step 4

Construct a wordsearch of at least 10 × 10 squares. You can make it as big as you like but keep it realistic. Work together with a group of friends. Many apps allow you to make wordsearch puzzles online. The words and phrases can go in any direction and phrases can be split. Your puzzle must only contain facts linked to the topic you are revising. Your task is to find 10 bits of information to hide in your puzzle, but you must not repeat information that you used in Step 3. DO NOT show where the words are. Fill up empty squares with random letters. Remember to keep a note of where your answers are hidden but do not show your friends. When you have a complete puzzle, exchange it with a friend to solve each other's puzzle.

Step 5

Now make up 10 questions (not "answers" this time) based on the same chapter used in the previous two tasks. Again, you must find NEW information that you have not yet used. Now it's getting hard to find that new information! Again, give your questions to a friend to answer.

Step 6

As you have been doing the puzzles, your brain has been actively searching for new information. Now write a NEW LIST that contains only the new information you have discovered when doing the puzzles. Your new list is the one to look at repeatedly for short bursts over the next few days. Try to remember more and more of it without looking at it. After a few days, you should be able to add words from your second list to your first list as you increase the information in your long-term memory.

FINALLY! Be inspired...

Make a list of different revision ideas and beside each one write **THINGS I HAVE** tried, **THINGS I WILL** try and **THINGS I MIGHT** try. Don't be scared of trying something new.

And remember – "FAIL TO PREPARE AND PREPARE TO FAIL!"

National 5 Physics

The exam

Duration: **2 hours**
Total marks: **110**

20 marks are awarded for 20 **multiple-choice questions** – completed on an answer grid.

90 marks are awarded for **written answers** – completed in the space provided after each question or on graph paper.

Approximately one third of the 110 marks are allocated to questions from each unit.

The National 5 Physics course consists of **three units**:

- Unit 1 – Electricity and Energy
- Unit 2 – Waves and Radiation
- Unit 3 – Dynamics and Space

General exam advice

There are 110 marks in total, and you have two hours to complete the paper. This works out at just over one minute per mark, so a 10 mark question would take roughly 11 minutes.

Be aware of how much time you spend on each question. For example, DO NOT spend 10 minutes on a question worth only three marks, especially when you haven't completed the rest of the questions – you can always return to the question later if there's time.

The best method for getting used to National 5 exam questions is to attempt as many exam type questions as possible, **and check your answers**. If you find a wrong answer, **find out why it is wrong** and then try similar questions until you can answer them correctly.

Specific exam advice

Advice for answering multiple-choice questions (Section 1) (20 marks)

Each question has five possible choices of answers. **Only one answer is correct.**

Multiple-choice questions are designed to test a range of skills, e.g.

- knowledge and understanding of the course
- using equations
- selecting correct statements from a list
- selecting and analysing information from a diagram.

It is important to **practise** as many **multiple-choice questions** as possible, to get used to the "style" and types of questions.

Do not try to work out all of the answers to multiple-choice questions in your head. Instead, when the question is complicated, write down notes and work on scrap paper (provided by the invigilator) or use the blank pages at the end of the question paper.

Do not use the answer grid for working, and remember to cross out your multiple-choice rough working when you have finished.

You can also make notes beside the actual question, if this helps, but **not** on the answer grid.

Advice for answering written questions (Section 2) (90 marks)

These questions test several different skills.

The majority of these marks test your **knowledge and understanding** of the course.

There are also questions which test different skills, like selecting information, analysing information, predicting results, and commenting on experimental results.

There are usually around 12–14 questions in Section 2. There are different types of questions, which include:

- Questions testing your **knowledge of the course**, sometimes applied to particular applications. More than half of the 90 marks in Section 2 are for this type of question.
- Questions (usually a maximum of two) involving **physics content not in the course** but explained in the question, usually including an equation which you are asked to use with data.
- A question testing your **scientific reading skills**, where you will be asked about a scientific report or passage. The question might include a calculation.
- **"Open-ended" questions** (a maximum of 2 per exam, three marks each), which usually discuss a physics phenomenon and ask you to explain it using your knowledge of physics. You have to think about the issue and try to give a step by step answer – there may be more than one area of physics used to answer this type of question. These questions allow you to use your knowledge and problem-solving skills. Be careful not to spend longer than necessary on these three mark questons.
- Questions testing practical skills usually based on **practical or experimental work**, which may have tables of results or graphs (or both) which have to be used to obtain information needed to answer the question. You could be asked to identify a problem with the results, or to suggest an improvement to the experiment.

Things to remember when answering questions

Using equations

More than half of the total marks awarded in Section 2 are for being able to calculate answers using an equation (relationship) from the **"Relationship Sheet"** which is supplied with the exam paper.

These questions are usually worth three marks. To obtain the full three marks for these questions, your final answer must be correct.

There are three separate marks awarded for the stages of the working:

- Write down the correct equation needed to calculate the answer from the Relationship Sheet – **1 mark**.
- Show that the correct values are substituted into the equation – **1 mark**.
- Show the final answer, including the correct unit – **1 mark**.

If the unit is wrong or missing, you will lose the final mark!

Other important areas to remember and practise are:

Units

The units of measurement in the National 5 physics course are based on the International System of Units. Make sure that you use the correct unit following a calculation in your final answer.

Prefixes

A prefix produces a multiple of the unit in powers of ten, e.g. 10^{-6} is 0·000001. It is named "micro" and has the symbol "μ". Make sure to practise and get used to all prefixes.

Scientific notation

This is used in the exam to write very large or very small numbers, to avoid writing or using strings of numbers in an answer or calculation.

You need to be familiar with how to enter and use numbers in scientific notation on **your** calculator – make sure that you have used your calculator often before the exam to get used to it.

Significant figures

When calculating a value using an equation, take care not to give too many significant figures in the final answer. If there are intermediate steps in a calculation, you can keep numbers in your calculator which have too many significant figures. You should always round your answer to give no more than the smallest number of significant figures which appear in the data given in the question.

E.g. $\frac{42{\cdot}74}{2{\cdot}59} = 16{\cdot}5019305$

If the smallest number of significant figures relating to the data used from the question was three, then round this answer to 16·5.

Examples:

- 20 has 1 significant figure
- 40·0 has 3 significant figures
- 0·000604 has 3 significant figures
- $4{\cdot}30 \times 10^4$ has 3 significant figures
- 6200 has 2 significant figures

Good luck!

Remember that the rewards for passing National 5 Physics are well worth it! Your pass will help you get the future you want for yourself. In the exam, be confident in your own ability. If you are not sure how to answer a question, trust your instincts and just give it a go anyway. Keep calm and don't panic! GOOD LUCK!

NATIONAL 5

Model Paper 1

Whilst this Model Paper has been specially commissioned by Hodder Gibson for use as practice for the National 5 exams, the key reference documents remain the SQA Specimen Paper 2013 and the SQA Past Papers 2014 and 2015.

National
Qualifications
MODEL PAPER 1

Physics
Section 1 — Questions

Date — Not applicable

Duration — 2 hours

Instructions for completion of Section 1 are given on Page two of the question paper.

Record your answers on the grid on Page three of your answer booklet

Do NOT write in this booklet.

Before leaving the examination room you must give your answer booklet to the Invigilator. If you do not, you may lose ALL the marks for this paper.

DATA SHEET

Speed of light in materials

Material	*Speed* in $m\,s^{-1}$
Air	$3{\cdot}0 \times 10^8$
Carbon dioxide	$3{\cdot}0 \times 10^8$
Diamond	$1{\cdot}2 \times 10^8$
Glass	$2{\cdot}0 \times 10^8$
Glycerol	$2{\cdot}1 \times 10^8$
Water	$2{\cdot}3 \times 10^8$

Gravitational field strengths

	Gravitational field strength on the surface in $N\,kg^{-1}$
Earth	9·8
Jupiter	23
Mars	3·7
Mercury	3·7
Moon	1·6
Neptune	11
Saturn	9·0
Sun	270
Uranus	8·7
Venus	8·9

Specific latent heat of fusion of materials

Material	*Specific latent heat of fusion* in $J\,kg^{-1}$
Alcohol	$0{\cdot}99 \times 10^5$
Aluminium	$3{\cdot}95 \times 10^5$
Carbon dioxide	$1{\cdot}80 \times 10^5$
Copper	$2{\cdot}05 \times 10^5$
Iron	$2{\cdot}67 \times 10^5$
Lead	$0{\cdot}25 \times 10^5$
Water	$3{\cdot}34 \times 10^5$

Specific latent heat of vaporisation of materials

Material	*Specific latent heat of vaporisation* in $J\,kg^{-1}$
Alcohol	$11{\cdot}2 \times 10^5$
Carbon dioxide	$3{\cdot}77 \times 10^5$
Glycerol	$8{\cdot}30 \times 10^5$
Turpentine	$2{\cdot}90 \times 10^5$
Water	$22{\cdot}6 \times 10^5$

Speed of sound in materials

Material	*Speed* in $m\,s^{-1}$
Aluminium	5200
Air	340
Bone	4100
Carbon dioxide	270
Glycerol	1900
Muscle	1600
Steel	5200
Tissue	1500
Water	1500

Specific heat capacity of materials

Material	*Specific heat capacity* in $J\,kg^{-1}\,°C^{-1}$
Alcohol	2350
Aluminium	902
Copper	386
Glass	500
Ice	2100
Iron	480
Lead	128
Oil	2130
Water	4180

Melting and boiling points of materials

Material	*Melting point* in °C	*Boiling point* in °C
Alcohol	−98	65
Aluminium	660	2470
Copper	1077	2567
Glycerol	18	290
Lead	328	1737
Iron	1537	2737

Radiation weighting factors

Type of radiation	*Radiation weighting factor*
alpha	20
beta	1
fast neutrons	10
gamma	1
slow neutrons	3

SECTION 1

1. In the circuit shown below, R_1 and R_2 are two identical resistors connected in series across a 24 V supply.

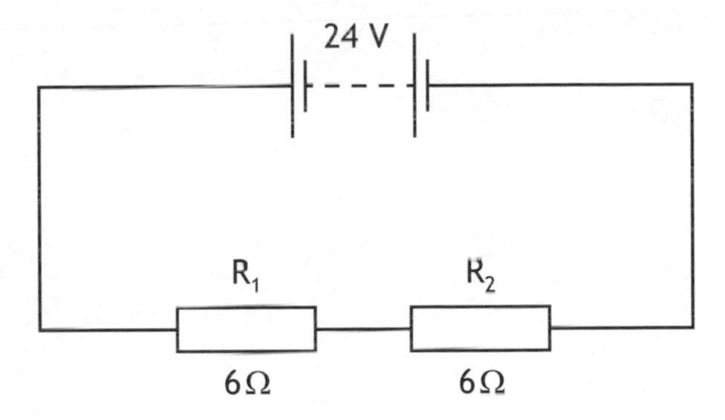

What is the current in resistor R_1?

A 0·25 A

B 0·5 A

C 2 A

D 4 A

E 8 A

2. Three resistors are connected as shown:

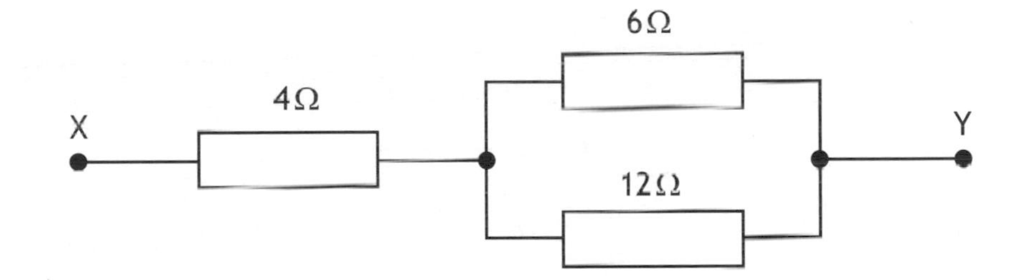

The total resistance between X and Y is

A 2 Ω

B 4 Ω

C 8 Ω

D 13 Ω

E 22 Ω.

3. The resistance of a wire is 6 Ω.

The current in the wire is 2A.

The power developed in the wire is

A 2W

B 3W

C 18W

D 24W

E 72W.

4. Which of the following is the correct symbol for a transistor?

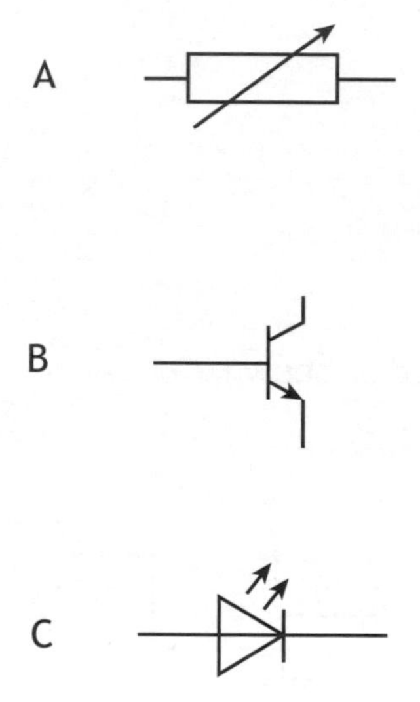

5. A student sets up the circuits shown.

In which circuit will both LEDs be lit?

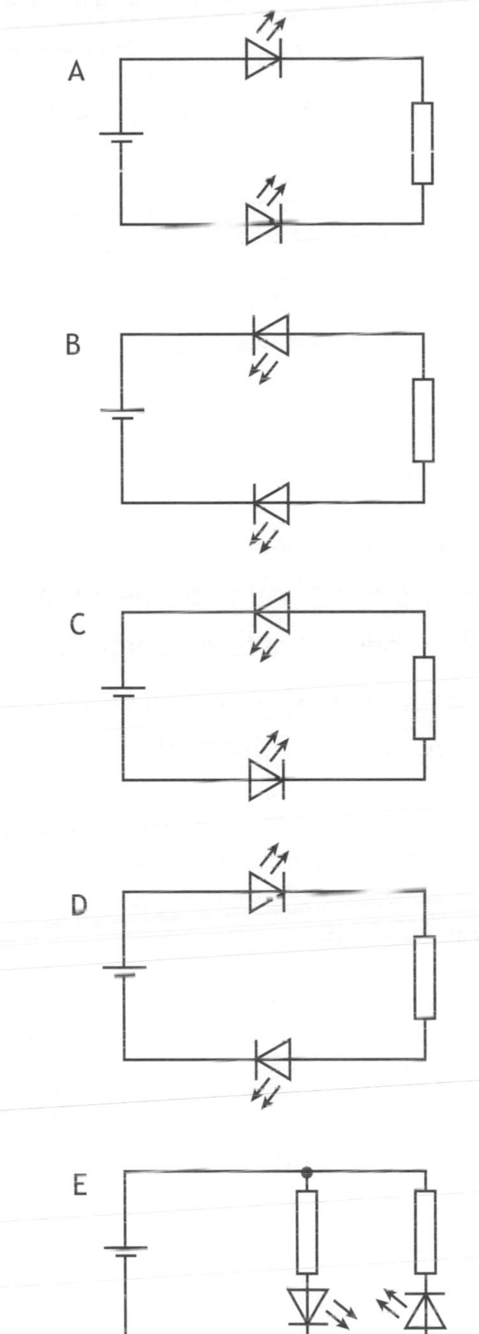

6. A passenger aircraft cruises at an altitude where the outside air pressure is $0{\cdot}35 \times 10^5$ Pa.

The air pressure inside the aircraft is $1{\cdot}0 \times 10^5$ Pa. The area of an external cabin door is $2{\cdot}2\ m^2$.

What is the outward force on the door due to the pressure difference?

A $0{\cdot}30 \times 10^5$ N

B $0{\cdot}61 \times 10^5$ N

C $1{\cdot}43 \times 10^5$ N

D $2{\cdot}2 \times 10^5$ N

E $2{\cdot}97 \times 10^5$ N.

7. For a fixed mass of gas at constant volume

A the pressure is directly proportional to temperature in K

B the pressure is inversely proportional to temperature in K

C the pressure is directly proportional to temperature in °C

D the pressure is inversely proportional to temperature in °C

E (pressure × temperature in K) is constant.

8. Ice at a temperature of −10 °C is heated until it becomes water at 80 °C.

The temperature change in kelvin is

A 70 K

B 90 K

C 343 K

D 363 K

E 636 K.

9. The energy of a water wave depends on its

A wavelength

B period

C colour

D speed

E amplitude.

10. A student can hear sound ranging from 20 Hz to 20 kHz.

If the velocity of sound in air is 340 $m s^{-1}$, the shortest wavelength the student can hear is

A 0·017 m

B 0·17 m

C 1·7 m

D 17 m

E 170 m.

11. The following diagram gives information about a wave.

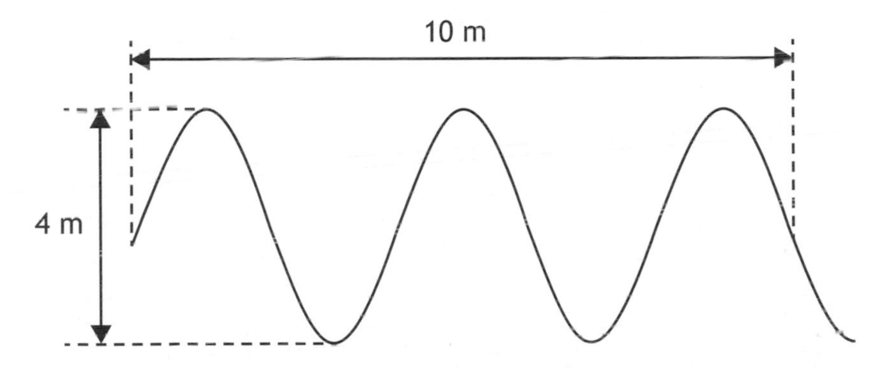

Which row shows the amplitude and wavelength of the wave?

	Amplitude (m)	*Wavelength* (m)
A	2	2
B	2	4
C	2	5
D	4	2
E	4	4

12. Which of the following lists the particles in order of size from smallest to largest?

A helium nucleus; electron; proton

B helium nucleus; proton; electron

C proton; helium nucleus; electron

D electron; helium nucleus; proton

E electron; proton; helium nucleus

13. A light wave has a frequency of $5 \cdot 2 \times 10^{14}$ Hz.

The period of the wave is

A $1 \cdot 9 \times 10^{-15}$ s

B $5 \cdot 8 \times 10^{-7}$ s

C $0 \cdot 19$ s

D $1 \cdot 7 \times 10^{6}$ s

E $1 \cdot 9 \times 10^{15}$ s.

14. A student makes the following statements for a ray of light travelling from glass into air.

I The direction of light always changes.

II The direction of light sometimes changes.

III The speed of light always changes.

IV The speed of light sometimes changes.

Which of these statements is/are correct?

A I and III only

B II and III only

C I and IV only

D III only

E IV only

15. A student makes the following statements about ionising radiations.

I Ionisation occurs when an atom loses an electron.

II Gamma radiation produces greater ionisation (density) than alpha particles.

III An alpha particle consists of 2 protons, 2 neutrons and 2 electrons.

Which of these statements is/are correct?

A I only

B II only

C I and II only

D II and III only

E I, II and III

16. Which of the following contains two scalar quantities?

A Weight and mass

B Distance and speed

C Force and mass

D Displacement and velocity

E Force and mass

17. The graph shows how the velocity of a ball changes with time.

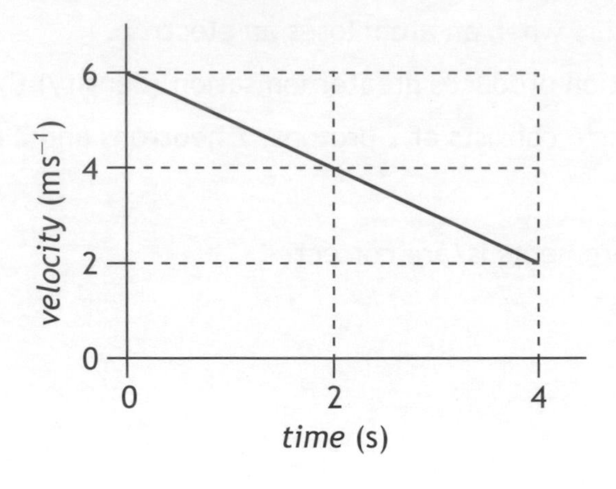

The acceleration of the ball is

A $-8\,ms^{-2}$

B $8\,ms^{-2}$

C $24\,ms^{-2}$

D $-1\,ms^{-2}$

E $1\,ms^{-2}$.

18. The Apollo Lunar Rover vehicle used by astronauts on the Moon had a mass of 204 kg.

Which row of the table gives the mass and weight of the vehicle on the moon?

	Mass (kg)	*Weight* (N)
A	55	55
B	55	204
C	204	204
D	204	326
E	204	1999

19. A ball is thrown horizontally from a cliff as shown.

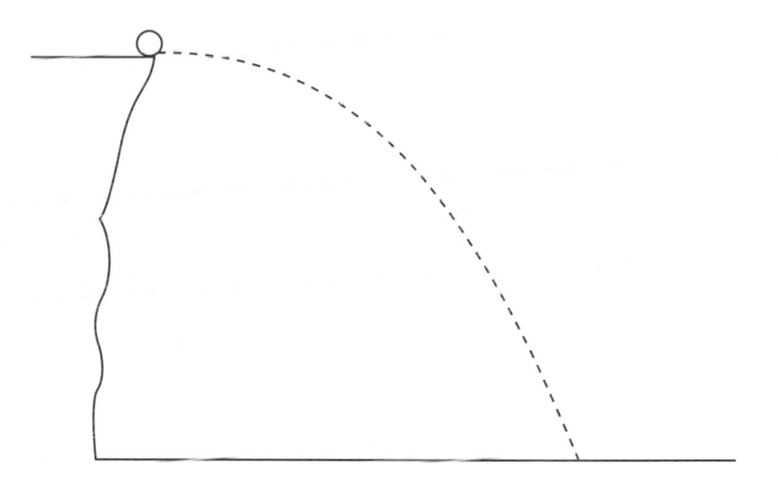

The effect of air resistance is negligible.

A student makes the following statements about the ball.

I The vertical speed of the ball increases as it falls.

II The vertical acceleration of the ball increases as it falls.

III The vertical force on the ball increases as it falls.

Which of these statements is/are correct?

A I only

B II only

C I and II only

D II and III only

E I, II and III

20. A rocket of mass 200 kg accelerates vertically upwards from the surface of a planet at $2 \cdot 0$ m s^{-2}.

The gravitational field strength on the planet is $4 \cdot 0$ N kg^{-1}.

What is the size of the force being exerted by the rocket's engines?

A 400 N

B 800 N

C 1200 N

D 2000 N

E 2400 N

[END OF SECTION 1. NOW ATTEMPT THE QUESTIONS IN SECTION 2 OF YOUR QUESTION AND ANSWER BOOKLET]

National
Qualifications
MODEL PAPER 1

Physics
Relationships Sheet

Date — Not applicable

$E_p = mgh$

$E_k = \frac{1}{2}mv^2$

$Q = It$

$V = IR$

$R_T = R_1 + R_2 + ...$

$\frac{1}{R_T} = \frac{1}{R_1} + \frac{1}{R_2} + ...$

$V_2 = \left(\frac{R_2}{R_1 + R_2}\right)V_s$

$\frac{V_1}{V_2} = \frac{R_1}{R_2}$

$P = \frac{E}{t}$

$P = IV$

$P = I^2R$

$P = \frac{V^2}{R}$

$E_h = cm\Delta T$

$p = \frac{F}{A}$

$\frac{pV}{T} = \text{constant}$

$p_1V_1 = p_2V_2$

$\frac{p_1}{T_1} = \frac{p_2}{T_2}$

$\frac{V_1}{T_1} = \frac{V_2}{T_2}$

$d = vt$

$v = f\lambda$

$T = \frac{1}{f}$

$A = \frac{N}{t}$

$D = \frac{E}{m}$

$H = Dw_R$

$\dot{H} = \frac{H}{t}$

$s = vt$

$d = \bar{v}t$

$s = \bar{v}t$

$a = \frac{v-u}{t}$

$W = mg$

$F = ma$

$E_w = Fd$

$E_h = ml$

[END OF SPECIMEN RELATIONSHIPS SHEET]

National
Qualifications
MODEL PAPER 1

Physics Section 1— Answer Grid and Section 2

Duration — 2 hours

Total marks — 110

SECTION 1 — 20 marks

Attempt ALL questions in this section.

Instructions for completion of Section 1 are given on Page two.

SECTION 2 — 90 marks

Attempt ALL questions in this section.

Read all questions carefully before answering.

Use **blue** or **black** ink. Do NOT use gel pens.

Write your answers in the spaces provided. Additional space for answers and rough work is provided at the end of this booklet. If you use this space, write clearly the number of the question you are answering. Any rough work must be written in this booklet. You should score through your rough work when you have written your fair copy.

Before leaving the examination room you must give this booklet to the Invigilator. If you do not, you may lose all the marks for this paper.

SECTION 1 — 20 marks

The questions for Section 1 are contained in the booklet Physics Section 1 — Questions. Read these and record your answers on the grid on Page three opposite.

1. The answer to each question is **either** A, B, C, D or E. Decide what your answer is, then fill in the appropriate bubble (see sample question below).

2. There is **only one correct** answer to each question.

3. Any rough working should be done on the rough working sheet.

Sample Question

The energy unit measured by the electricity meter in your home is the:

A ampere

B kilowatt-hour

C watt

D coulomb

E volt.

The correct answer is **B**—kilowatt-hour. The answer **B** bubble has been clearly filled in (see below).

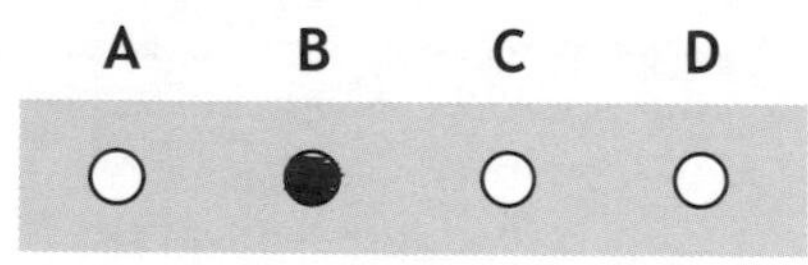

Changing an answer

If you decide to change your answer, cancel your first answer by putting a cross through it (see below) and fill in the answer you want. The answer below has been changed to **D**.

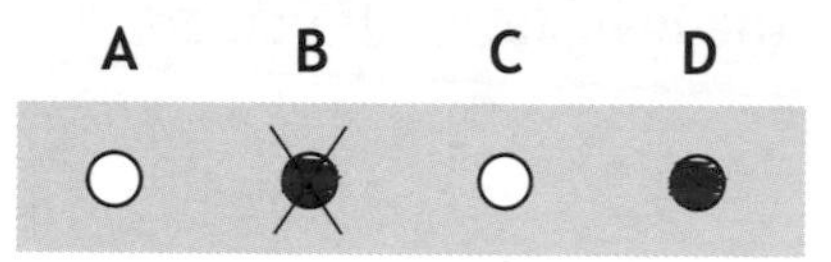

If you then decide to change back to an answer you have already scored out, put a tick (✓) to the **right** of the answer you want, as shown below:

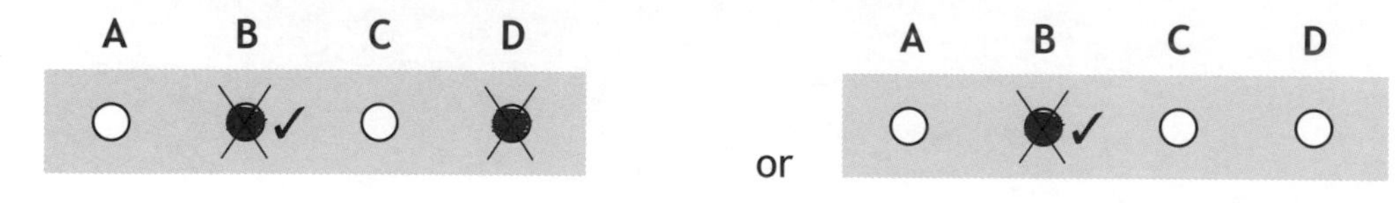

SECTION 1 — Answer Grid

	A	B	C	D	E
1	○	○	○	○	○
2	○	○	○	○	○
3	○	○	○	○	○
4	○	○	○	○	○
5	○	○	○	○	○
6	○	○	○	○	○
7	○	○	○	○	○
8	○	○	○	○	○
9	○	○	○	○	○
10	○	○	○	○	○
11	○	○	○	○	○
12	○	○	○	○	○
13	○	○	○	○	○
14	○	○	○	○	○
15	○	○	○	○	○
16	○	○	○	○	○
17	○	○	○	○	○
18	○	○	○	○	○
19	○	○	○	○	○
20	○	○	○	○	○

[BLANK PAGE]

SECTION 2 — 90 marks

Attempt ALL questions

MARKS | DO NOT WRITE IN THIS MARGIN

1. While repairing a school roof, workmen lift a pallet of tiles from the ground to the top of the scaffolding.

This job is carried out using a motorised pulley system.

The pallet and tiles have a total mass of 235 kg.

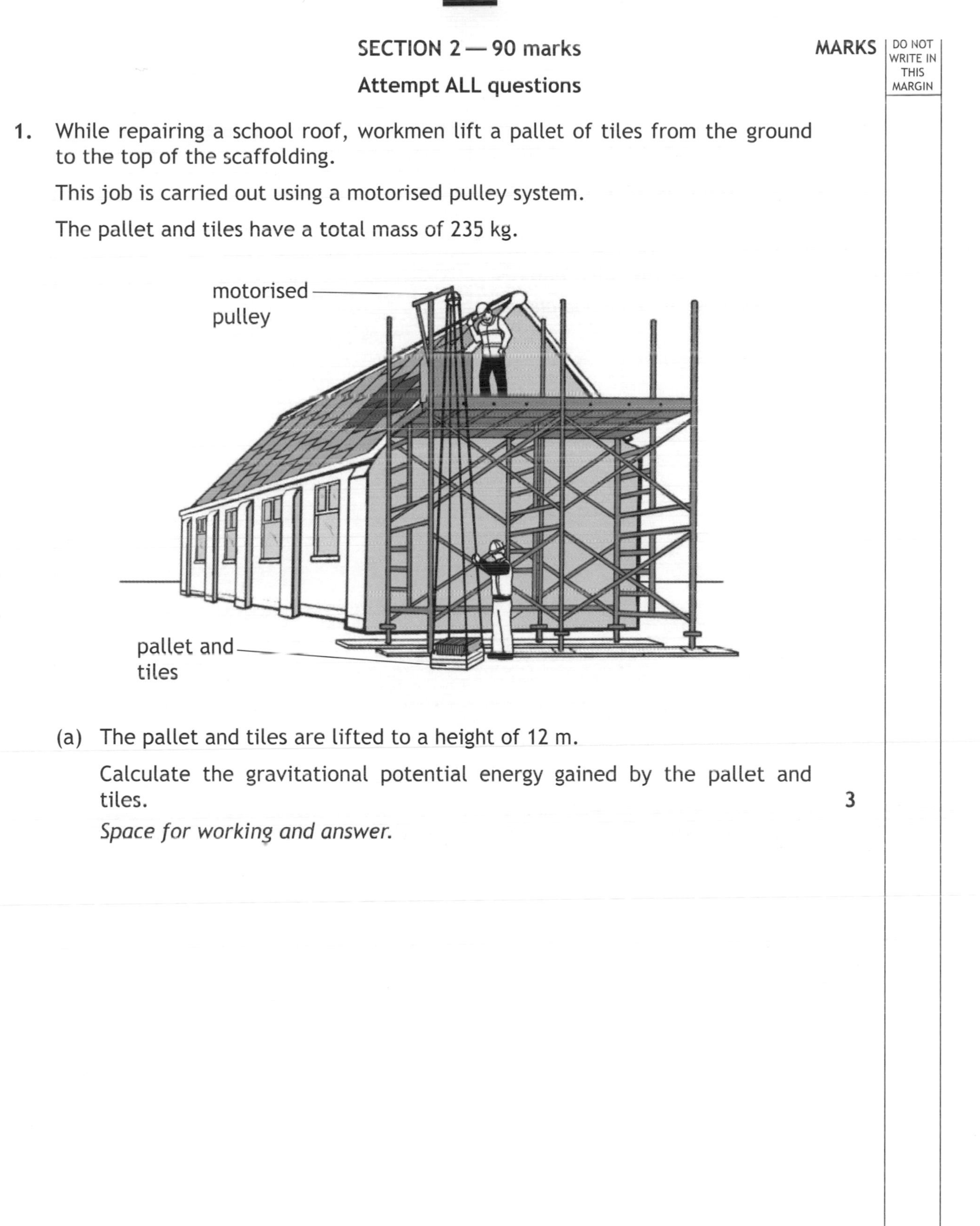

(a) The pallet and tiles are lifted to a height of 12 m.

Calculate the gravitational potential energy gained by the pallet and tiles. **3**

Space for working and answer.

MARKS | DO NOT WRITE IN THIS MARGIN

1. (continued)

(b) When the tiles are being unloaded onto the scaffolding, at a height of 12 m, one tile falls.

The tile has a mass of 2·5 kg.

(i) Calculate the final speed of the tile just before it hits the ground.

Assume the tile falls from rest. **4**

Space for working and answer.

(ii) Explain why the actual speed is less than the speed calculated in (b)(i). **1**

Total marks 8

MARKS | DO NOT WRITE IN THIS MARGIN

2. A student sets up the following circuit to investigate the resistance of resistor R.

The variable resistor is adjusted and the voltmeter and ammeter readings are noted. The following graph is obtained from the experimental results.

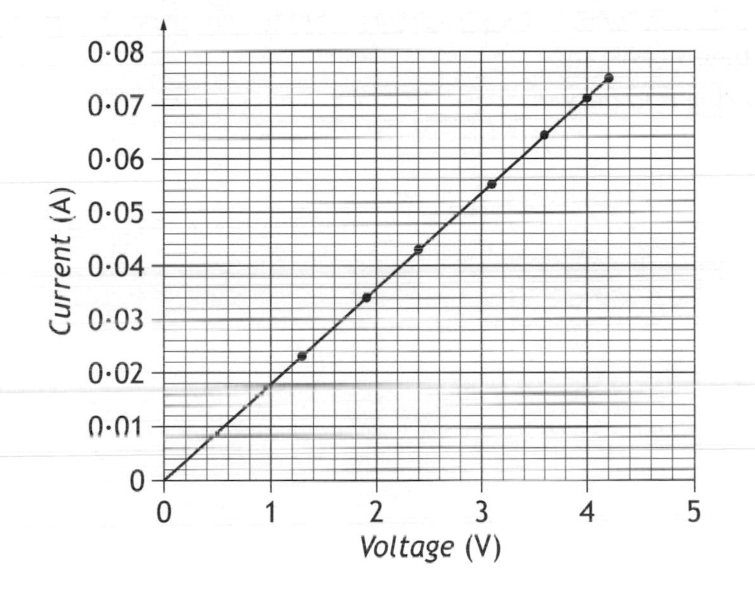

(a) (i) Calculate the value of the resistor R when the reading on the voltmeter is 4·2 V. **3**

Space for working and answer.

MARKS DO NOT WRITE IN THIS MARGIN

2. (continued)

(ii) Using information from the graph, state whether the resistance of the resistor R, increases, stays the same or decreases as the voltage increases.

You must justify your answer. **2**

(b) The student is given a task to combine two resistors from a pack containing one each of 33 Ω, 56 Ω, 82 Ω, 150 Ω, 270 Ω, 390 Ω.

Show by calculation which two resistors should be used to give the smallest combined resistance. **3**

Space for working and answer.

Total marks 8

MARKS DO NOT WRITE IN THIS MARGIN

3. An important experiment which resulted in determining the value for the charge on an electron was carried out by a scientist named Millikan in 1909.

Part of the experiment required the **terminal velocity** of tiny drops of oil to be determined as they fell through air.

(a) Explain what is meant by terminal velocity. 1

(b) As the oil drop fell through the air, Milliken used an equation known as Stokes' Law to determine the upward drag force, F_d, acting on the drop:

$$F_d = 6\pi r \eta v_1$$

Where:

v_1 is the terminal velocity of the falling drop,

η is the viscosity of the air,

r is the radius of the drop

For one particular oil drop, its radius was 2·83 μm, its terminal velocity was $8{\cdot}56 \times 10^{-4}$ m s^{-1} and the viscosity of air was $1{\cdot}820 \times 10^{-5}$ kg m^{-1} s^{-1}.

Calculate the value of the drag force, F_d, acting on the oil drop. 2

Space for working and answer.

MARKS | DO NOT WRITE IN THIS MARGIN

3. (continued)

(c) The weight of another oil drop was $8{\cdot}6 \times 10^{-13}$ N.

Calculate the mass of this oil drop. **3**

Space for working and answer.

(d) For another part of the experiment, a quantity of charge was added to the oil drop. The drop was then placed in an electric field.

State the effect of an electric field on a charged particle. **1**

Total marks 7

MARKS DO NOT WRITE IN THIS MARGIN

4. A solar furnace consists of an array of mirrors which reflect heat radiation on to a central curved reflector.

A heating container is placed at the focus of the central curved reflector.

Metals placed in the container are heated until they melt.

(a) 8000 kg of pre-heated aluminium pellets at a temperature of 160 °C are placed in the container. Aluminium has a specific heat capacity of 902 J $kg^{-1}\,°C^{-1}$ and a melting point of 660 °C.

Calculate the heat energy required to heat the aluminium to its melting point. **3**

Space for working and answer.

(b) How much extra energy is required to melt the aluminium pellets? **3**

Space for working and answer.

Total marks 6

MARKS | DO NOT WRITE IN THIS MARGIN

5. Estimate the pressure exerted on the floor by an average National 5 student who is standing on two feet.

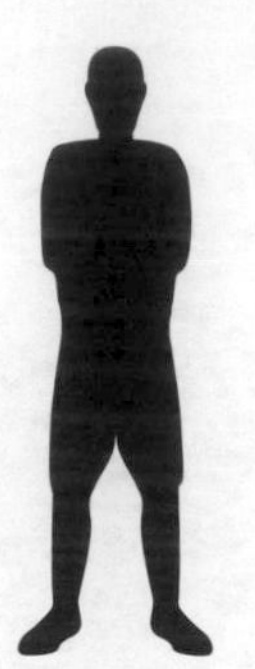

Show any working clearly, and explain any assumptions that you make. **3**

MARKS | DO NOT WRITE IN THIS MARGIN

6. A student is training to become a diver.

The student carries out an experiment to investigate the relationship between the pressure and volume of a fixed mass of gas using the apparatus shown.

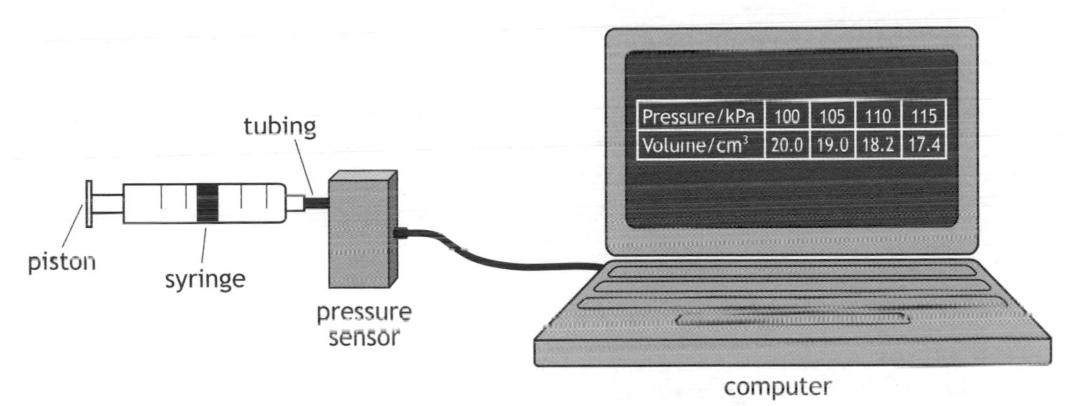

The pressure of the gas is recorded using a pressure sensor connected to a computer. The volume of the gas is also recorded. The student pushes the piston to alter the volume and a series of readings is taken.

The temperature of the gas is constant during the experiment.

The results are shown below.

Pressure (kPa)	100	105	110	115
Volume (cm^3)	20·0	19·0	18·2	17·4

(a) Using all the data, establish the relationship between the pressure and volume of the gas. **2**

(b) Use the kinetic model to explain the change in pressure as the volume of gas decreases. **3**

(c) Explain why it is important for the tubing to be as short as possible. **1**

Total marks 6

MARKS | DO NOT WRITE IN THIS MARGIN

7. (a) Refraction of light occurs in glass.

What is meant by the term refraction? **1**

(b) The following diagram shows a ray of light entering a glass block.

(i) Complete the diagram to show the path of the ray of light through the block and after it emerges from the block. **2**

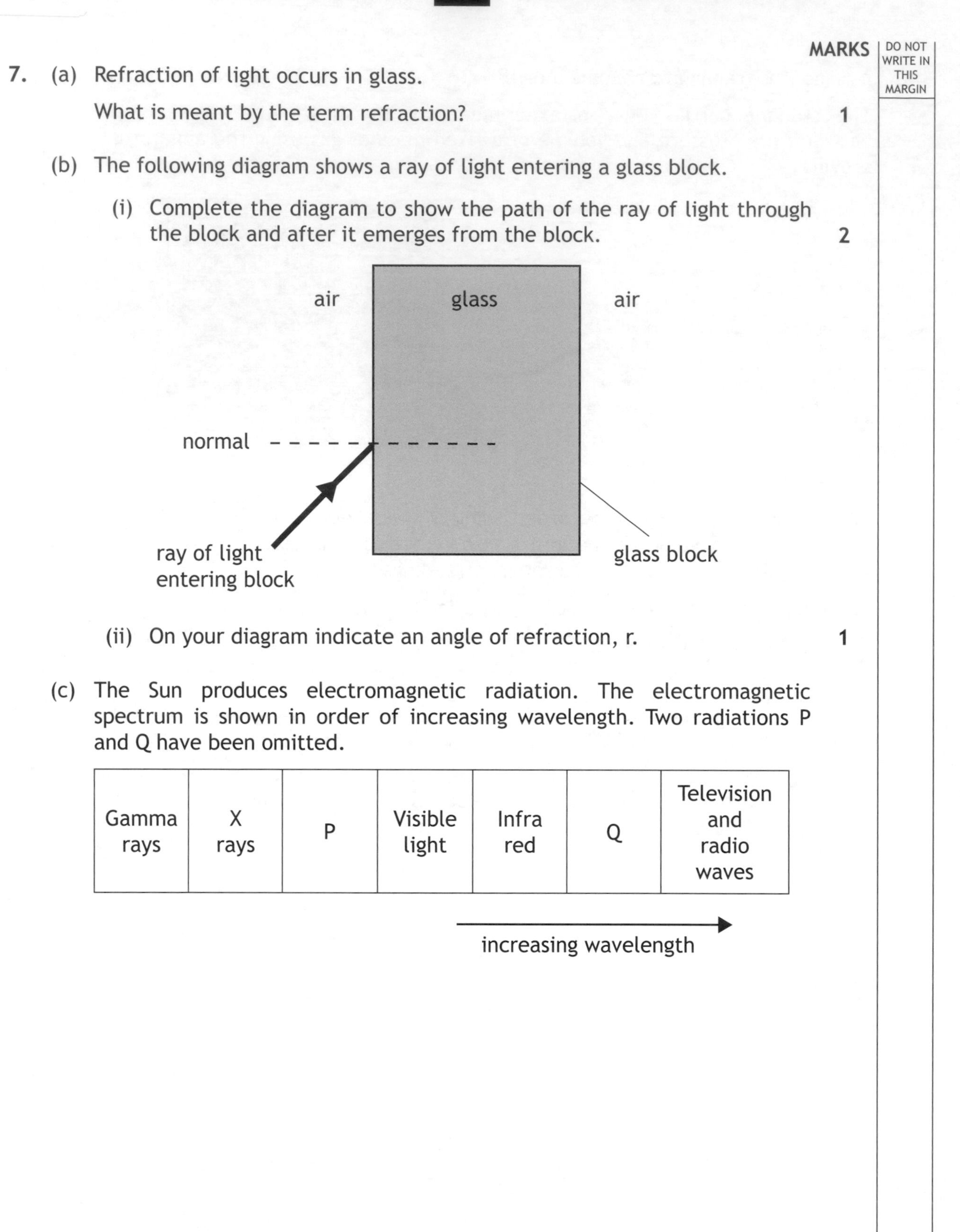

(ii) On your diagram indicate an angle of refraction, r. **1**

(c) The Sun produces electromagnetic radiation. The electromagnetic spectrum is shown in order of increasing wavelength. Two radiations P and Q have been omitted.

Gamma rays	X rays	P	Visible light	Infra red	Q	Television and radio waves

increasing wavelength →

MARKS DO NOT WRITE IN THIS MARGIN

7. (continued)

(i) Identify radiations P and Q. **2**

(ii) The planet Neptune is $4{\cdot}50 \times 10^9$ km from the Sun.

Calculate the time taken for radio waves from the Sun to reach Neptune. **3**

Space for working and answer.

Total marks 9

MARKS DO NOT WRITE IN THIS MARGIN

8. In 1908 Ernest Rutherford conducted a series of experiments involving alpha particles.

(a) State what is meant by an alpha particle. **1**

(b) Alpha particles produce a greater ionisation density than beta particles or gamma rays.

What is meant by the term *ionisation*? **1**

MARKS | DO NOT WRITE IN THIS MARGIN

8. (continued)

(c) A radioactive source emits alpha particles and has a half-life of 2·5 hours.

The source has an initial activity of 4·8 kBq.

Calculate the time taken for its activity to decrease to 300 Bq. **3**

Space for working and answer.

(d) Some sources emit alpha particles and are stored in lead cases despite the fact that alpha particles cannot penetrate paper.

Suggest a possible reason for storing these sources using this method. **1**

Total marks 6

MARKS | DO NOT WRITE IN THIS MARGIN

9. An ageing nuclear power station is being dismantled.

(a) During the dismantling process a worker comes into contact with an object that emits 24000 alpha particles in five minutes. The worker's hand has a mass of 0·50 kg and absorbs 6·0 μJ of energy.

(i) Calculate the absorbed dose received by the worker's hand. **3**

Space for working and answer.

MARKS | DO NOT WRITE IN THIS MARGIN

9. (continued)

(ii) Calculate the equivalent dose received by the worker's hand. **3**

Space for working and answer.

(iii) Calculate the activity of the object. **3**

Space for working and answer.

(b) What type of nuclear reaction takes place in a nuclear power station's reactor? **1**

Total marks 10

MARKS DO NOT WRITE IN THIS MARGIN

10. Two cyclists choose different routes to travel from point **A** to a point **B** some distance away.

(a) Cyclist X travels 12 km due east (090).

He then turns and travels due south (180) and travels a further 15 km to arrive at **B**.

He takes 1 hour 15 minutes to travel from **A** to **B**.

(i) By scale drawing (or otherwise) find the displacement of **B** from **A**. **4**

MARKS | DO NOT WRITE IN THIS MARGIN

10. (continued)

(ii) Calculate the average velocity of cyclist X for the journey from **A** to **B**. **3**

Space for working and answer.

(b) Cyclist Y travels a total distance of 33 km by following a different route from **A** to **B** at an average speed of 22 $km\,h^{-1}$.

State the displacement of cyclist Y on completing this route. **1**

Total marks 8

MARKS DO NOT WRITE IN THIS MARGIN

11. A child sledges down a hill.

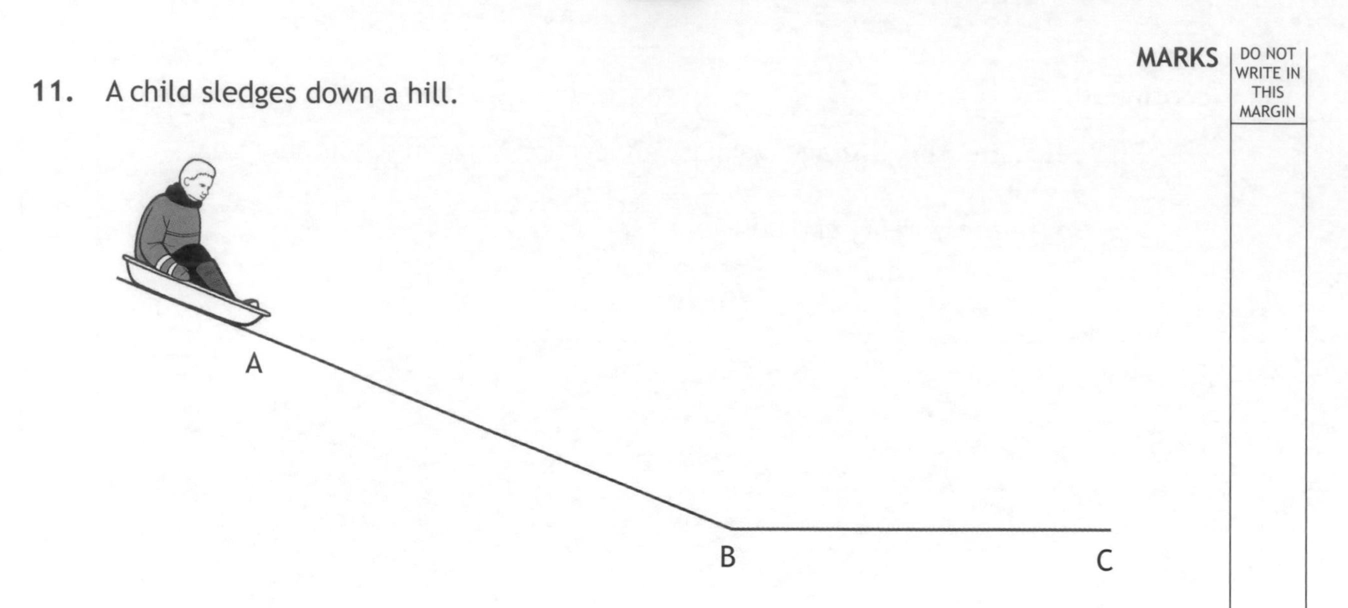

The sledge and child are released from rest at point A. They reach a speed of 3 m s^{-1} at point B.

(a) The sledge and child take 5s to reach point B.

Calculate the acceleration. **3**

Space for working and answer.

(b) The sledge and child have a combined mass of 40 kg.

Calculate the unbalanced force acting on them. **3**

Space for working and answer.

Total marks 6

MARKS DO NOT WRITE IN THIS MARGIN

12. An underwater generator is designed to produce electricity from water currents in the sea.

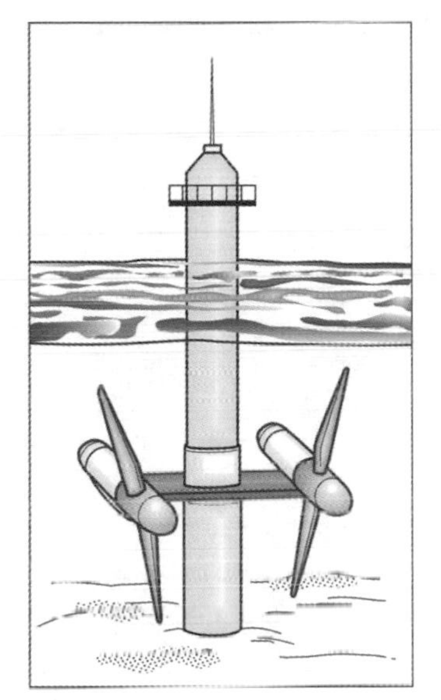

The output power of the generator depends on the speed of the water current as shown in Graph 1.

Graph 1

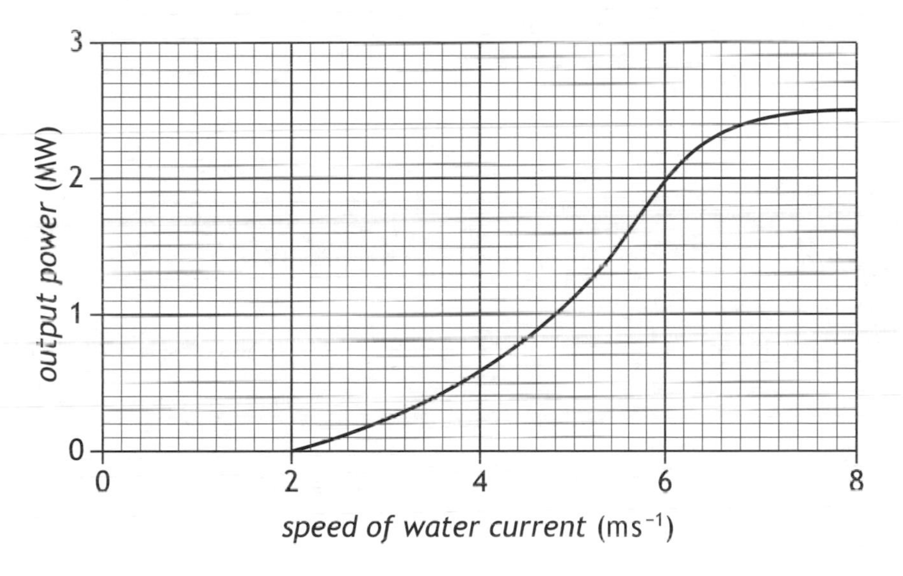

The speed of the water current is recorded at different times of the day shown in Graph 2.

Graph 2

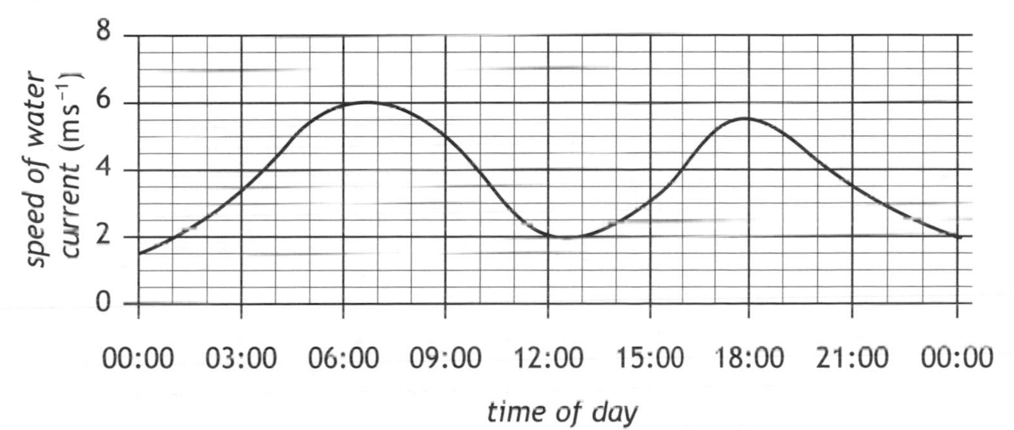

MARKS | DO NOT WRITE IN THIS MARGIN

12. (continued)

(a) (i) State the output power of the generator at 09:00. **1**

(ii) State one disadvantage of using this type of generator. **1**

(b) Three different types of electrical generator, X, Y and Z are tested in a special tank with a current of water as shown to find out the efficiency of each generator.

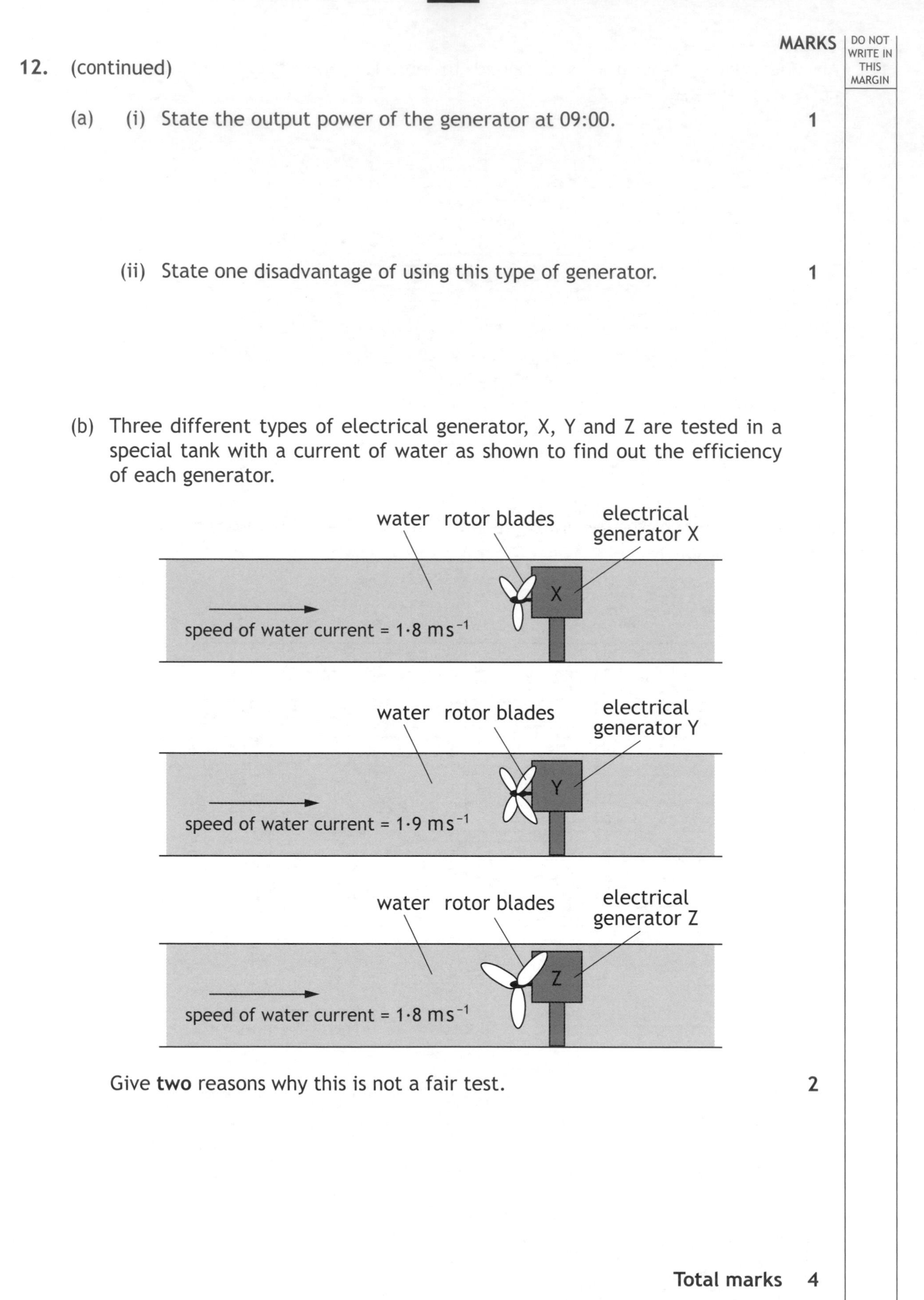

Give **two** reasons why this is not a fair test. **2**

Total marks 4

MARKS DO NOT WRITE IN THIS MARGIN

13. Read the passage below and answer the questions that follow.

Black holes

A black hole is where the force of gravity is so strong because matter has collapsed into a tiny space. This can happen when a star has used up most of its energy.

Because no light can get out, black holes cannot be seen. Space telescopes can help find black holes. Using special instruments on satellites, the behaviour of stars which are close to black holes can be monitored.

Black holes can have a range of sizes. A "stellar" black hole has a mass of up to 20 times more than the mass of the sun. There may be many stellar mass black holes in the Milky Way galaxy.

There is evidence which suggests that at the centre of the Milky Way there is a supermassive black hole.

A supermassive black hole has a mass of more than 1 million suns. Scientists have discovered that every large galaxy contains a supermassive black hole at its centre.

At a distance of $2{\cdot}6 \times 10^{20}$ metres from Earth, the supermassive black hole at the centre of the Milky Way is called Sagittarius A. Sagittarius A would fit inside a large sphere which could hold several million Earth masses.

Scientists think the supermassive black holes formed when the universe began.

Stellar black holes are made when the centre of a very big star collapses inwards on itself. When this happens, it causes a supernova. A supernova is an exploding star that blasts part of the star into space.

Black holes cannot be seen directly because gravity prevents light escaping from the black hole. Observation of how nearby stars are affected by their strong gravity provides information about the behaviour, size and nature of the black hole.

The interaction of stars and black holes when they are close together produces intense gamma radiation. Satellites and telescopes in space are used to detect this radiation.

MARKS | DO NOT WRITE IN THIS MARGIN

13. (continued)

(a) Name a black hole mentioned in the passage. **1**

(b) Calculate how many light years the Earth is from the centre of the Milky Way. **3**

Space for working and answer.

(c) Telescopes on satellites are used to detect light rays and gamma radiation.

(i) Name a detector of gamma rays. **1**

(ii) Complete the sentences by circling the correct words. **1**

Compared to gamma rays, light rays have a { higher / lower } frequency which means they have a { higher / lower } energy.

Total marks 6

MARKS | DO NOT WRITE IN THIS MARGIN

14. What affects how long it takes objects fall to the ground?

Use your knowledge of physics to answer this question. **3**

[END OF MODEL PAPER]

MARKS DO NOT WRITE IN THIS MARGIN

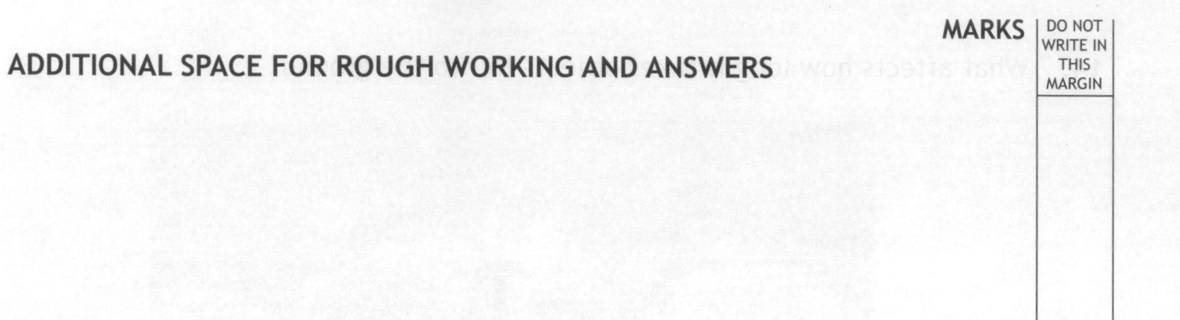

ADDITIONAL SPACE FOR ROUGH WORKING AND ANSWERS

MARKS DO NOT WRITE IN THIS MARGIN

ADDITIONAL SPACE FOR ROUGH WORKING AND ANSWERS

MARKS DO NOT WRITE IN THIS MARGIN

ADDITIONAL SPACE FOR ROUGH WORKING AND ANSWERS

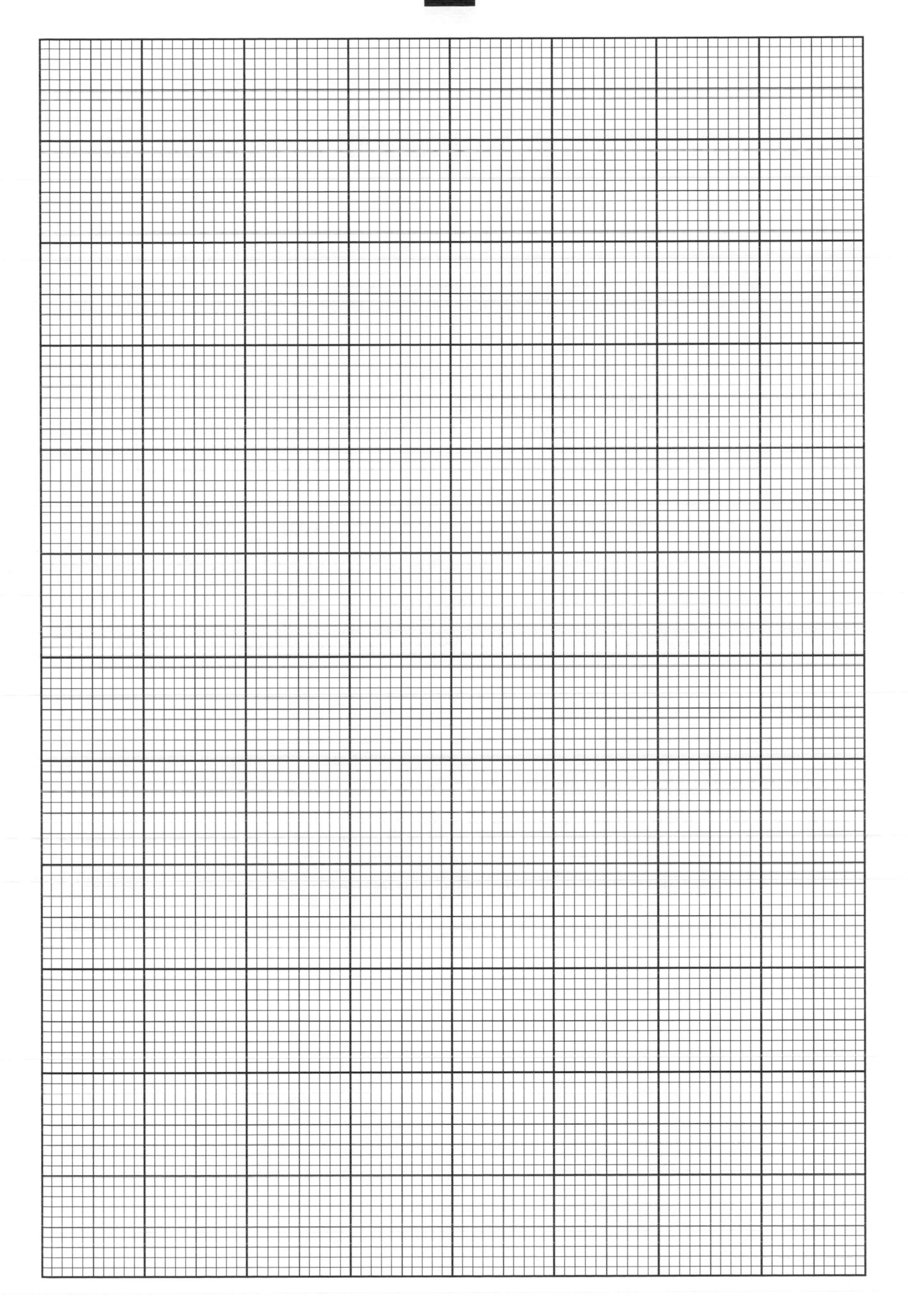

[BLANK PAGE]

NATIONAL 5

Model Paper 2

Whilst this Model Paper has been specially commissioned by Hodder Gibson for use as practice for the National 5 exams, the key reference documents remain the SQA Specimen Paper 2013 and the SQA Past Papers 2014 and 2015.

National
Qualifications
MODEL PAPER 2

Physics
Section 1—Questions

Date — Not applicable

Duration — 2 hours

Instructions for completion of Section 1 are given on Page two of the question paper.

Record your answers on the grid on Page three of your answer booklet

Do NOT write in this booklet.

Before leaving the examination room you must give your answer booklet to the Invigilator.
If you do not, you may lose ALL the marks for this paper.

DATA SHEET

Speed of light in materials

Material	*Speed* in m s^{-1}
Air	$3{\cdot}0 \times 10^8$
Carbon dioxide	$3{\cdot}0 \times 10^8$
Diamond	$1{\cdot}2 \times 10^8$
Glass	$2{\cdot}0 \times 10^8$
Glycerol	$2{\cdot}1 \times 10^8$
Water	$2{\cdot}3 \times 10^8$

Speed of sound in materials

Material	*Speed* in m s^{-1}
Aluminium	5200
Air	340
Bone	4100
Carbon dioxide	270
Glycerol	1900
Muscle	1600
Steel	5200
Tissue	1500
Water	1500

Gravitational field strengths

	Gravitational field strength on the surface in N kg^{-1}
Earth	9·8
Jupiter	23
Mars	3·7
Mercury	3·7
Moon	1·6
Neptune	11
Saturn	9·0
Sun	270
Uranus	8·7
Venus	8·9

Specific heat capacity of materials

Material	*Specific heat capacity* in J kg^{-1} °C^{-1}
Alcohol	2350
Aluminium	902
Copper	386
Glass	500
Ice	2100
Iron	480
Lead	128
Oil	2130
Water	4180

Specific latent heat of fusion of materials

Material	*Specific latent heat of fusion* in J kg^{-1}
Alcohol	$0{\cdot}99 \times 10^5$
Aluminium	$3{\cdot}95 \times 10^5$
Carbon dioxide	$1{\cdot}80 \times 10^5$
Copper	$2{\cdot}05 \times 10^5$
Iron	$2{\cdot}67 \times 10^5$
Lead	$0{\cdot}25 \times 10^5$
Water	$3{\cdot}34 \times 10^5$

Melting and boiling points of materials

Material	*Melting point* in °C	*Boiling point* in °C
Alcohol	−98	65
Aluminium	660	2470
Copper	1077	2567
Glycerol	18	290
Lead	328	1737
Iron	1537	2737

Specific latent heat of vaporisation of materials

Material	*Specific latent heat of vaporisation* in J kg^{-1}
Alcohol	$11{\cdot}2 \times 10^5$
Carbon dioxide	$3{\cdot}77 \times 10^5$
Glycerol	$8{\cdot}30 \times 10^5$
Turpentine	$2{\cdot}90 \times 10^5$
Water	$22{\cdot}6 \times 10^5$

Radiation weighting factors

Type of radiation	*Radiation weighting factor*
alpha	20
beta	1
fast neutrons	10
gamma	1
slow neutrons	3

SECTION 1

1. A student makes the following statements about electric fields.

   I There is a force on a charge in an electric field.

   II When an electric field is applied to a conductor, the free electric charges in the conductor move.

   III Work is done when a charge is moved in an electric field.

   Which of these statements is/are correct?

   A I only

   B II only

   C I and II only

   D I and III only

   E I, II and III

2. The circuit below shows a 20 V supply connected across two resistors.

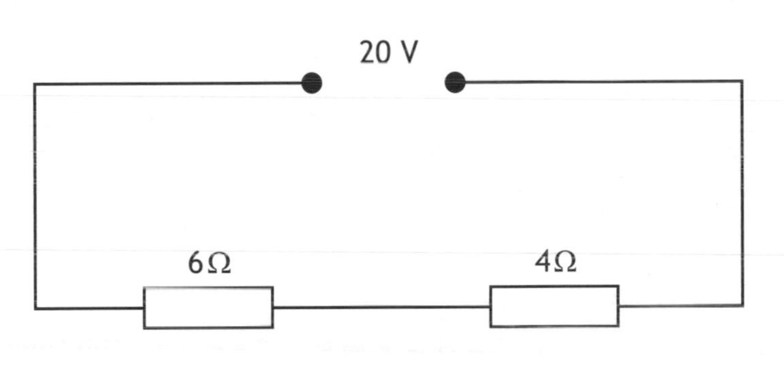

   The charge passing through the 6 Ω resistor in 3 s is

   A 2 C

   B 6 C

   C 8 C

   D 12 C

   E 20 C.

3. Three resistors are connected as shown.

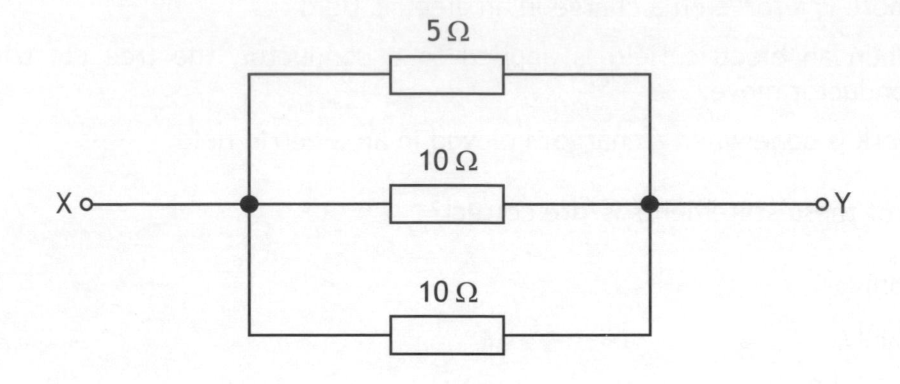

The resistance between X and Y is

A 0·06 Ω

B 0·4 Ω

C 2·5 Ω

D 13 Ω

E 25 Ω.

4. Resistors are connected in the following circuit as shown.

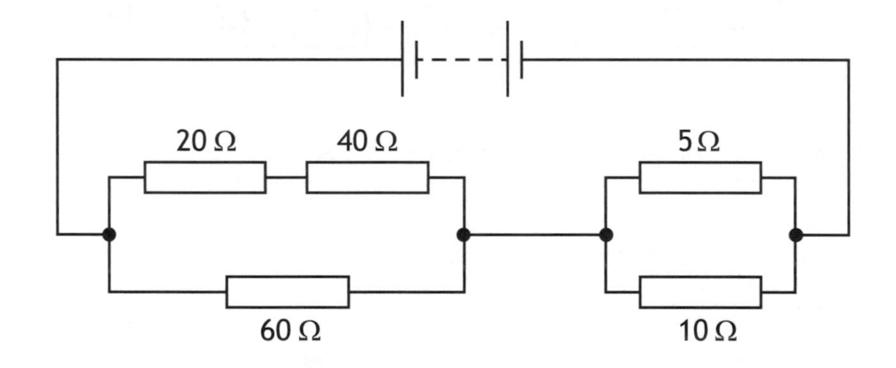

In which resistor is the current smallest?

A 5 Ω

B 10 Ω

C 20 Ω

D 40 Ω

E 60 Ω.

5. A car headlamp bulb has 24 W, 12 V printed on it.

What is the current in the bulb and the resistance of the filament when operating normally?

	Normal working current (A)	*Resistance of the filament* (Ω)
A	0·5	24
B	0·5	6
C	2	6
D	2	24
E	2	48

6. A temperature of 273 °C is the same as a temperature of

A 0 K

B 100 K

C 273 K

D 373 K

E 546 K.

7. The International Space Station satellite has a period of 103 minutes and an orbital height of 400 km.

The GEOS–15 metrological satellite has a period of 1436 minutes and an orbital height of 36 000 km.

Which of the following gives the period of the GLONASS global positioning satellite which has an orbital height of 19 000 km?

A 82 minutes

B 103 minutes

C 675 minutes

D 1436 minutes

E 1539 minutes.

8. Which of the following electromagnetic waves has a higher frequency than microwaves and a lower frequency than visible light?

A Gamma rays

B Infrared

C Radio

D Ultraviolet

E X-rays.

9. A ray of light passes from air into a glass block as shown.

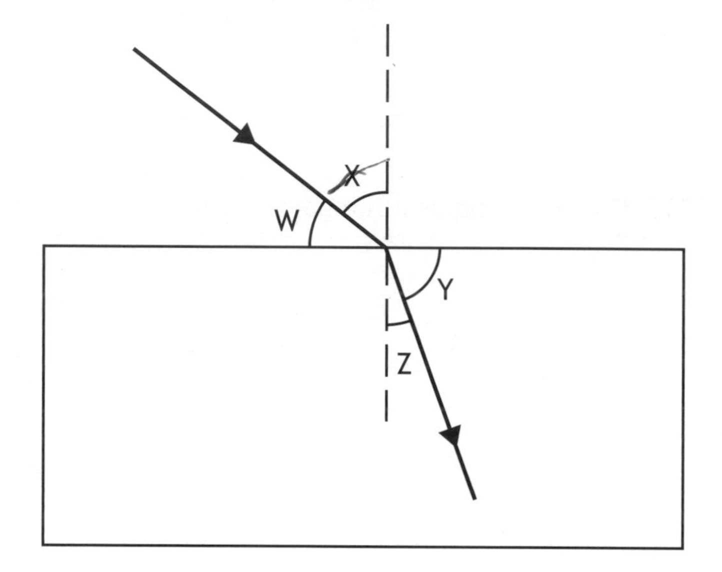

Which row in the table shows the angle of incidence and the angle of refraction?

	Angle of incidence	*Angle of refraction*
A	W	Z
B	W	Y
C	X	Z
D	X	Y
E	Z	X

10. A student makes the following statements.

I The nucleus of an atom contains protons and electrons.

II Gamma radiation produces the greatest ionisation density.

III Beta particles are fast moving electrons.

Which of these statements is/are correct?

A I only

B I and III only

C II only

D II and III only

E III only

11. A radioactive source emits alpha, beta and gamma radiation. A detector, connected to a counter, is placed 10 mm in front of the source. The counter records 400 counts per minute.

A sheet of paper is placed between the source and the detector. The counter records 300 counts per minute.

The radiation now detected is

A alpha only

B alpha and beta only

C beta only

D beta and gamma only

E gamma only.

12. A radioactive tracer is a liquid which is injected into a patient to study the flow of blood.

The radioactive tracer is carried around the body by the patient's blood.

A detector is placed above the patient to monitor the flow of blood carrying the tracer.

The tracer should have a

A short half-life and emit α particles

B long half-life and emit β particles

C long half-life and emit γ rays

D long half-life and emit α particles

E short half-life and emit γ rays.

13. For a particular radioactive source, 1800 atoms decay in a time of 3 minutes. The **activity** of this source is

A 10 Bq

B 600 Bq

C 800 Bq

D 5400 Bq

E 324 000 Bq.

14. Human tissue can be damaged by exposure to radiation.

On which of the following factors does the risk of biological harm depend?

I The absorbed dose.

II The type of radiation.

III The body organs or tissue exposed.

A I only

B I and II only

C II only

D II and III only

E I, II and III

15. At an airport an aircraft moves from the terminal building to the end of the runway.

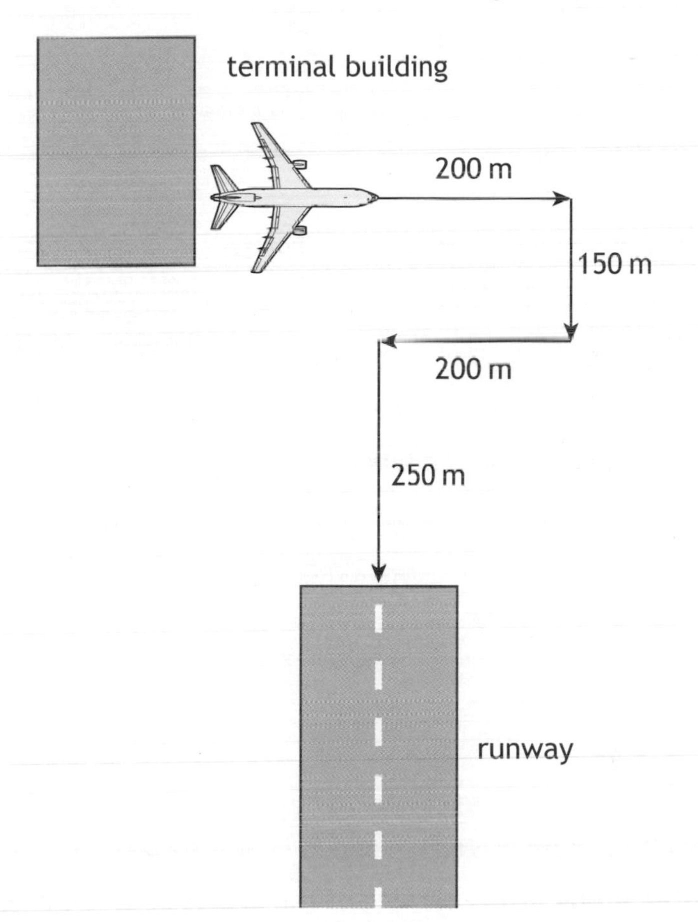

Which row shows the total distance travelled and the size of the displacement of the aircraft?

	Total distance travelled (m)	*Size of displacement* (m)
A	400	800
B	450	200
C	450	400
D	800	400
E	800	800

16. A block is pulled across a horizontal surface as shown.

The mass of the block is 6 kg.

The block is travelling at constant speed.

The force of friction acting on the block is

A 0 N

B 5 N

C 24 N

D 30 N

E 36 N.

17. A space probe with a mass of 760 kg landed on the surface of a planet in our solar system.

The weight of the probe at the surface of the planet in our solar system was 6764 N.

The planet was

A Jupiter

B Mars

C Neptune

D Saturn

E Venus.

18. Near the Earth's surface, a mass of 5 kg is falling with a constant velocity.

The air resistance and the unbalanced force acting on the mass are:

	air resistance	*unbalanced force*
A	49 N upwards	49 N downwards
B	9·8 N upwards	9·8 N downwards
C	9·8 N downwards	58·8 N downwards
D	9·8 N upwards	0 N
E	49 N upwards	0 N

19. Two identical balls X and Y are projected horizontally from the edge of a cliff. The path taken by each ball is shown.

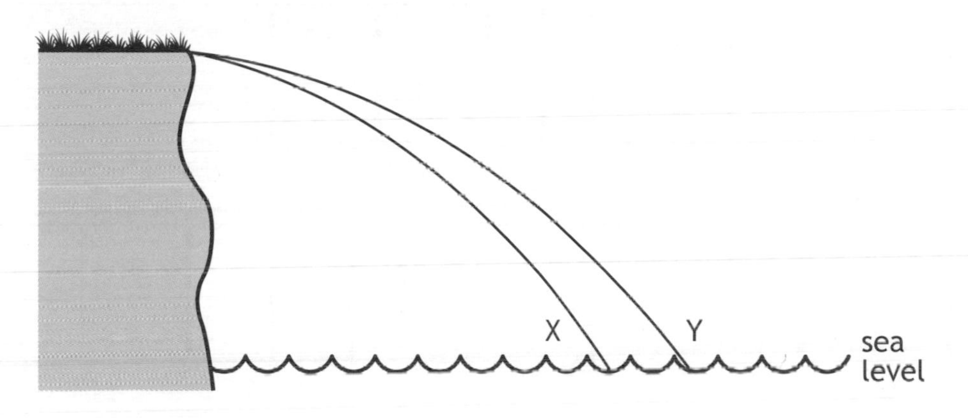

A student makes the following statements about the motion of the two balls.

I They take the same time to reach sea level.
II They have the same vertical acceleration.
III They have the same horizontal velocity.

Which of these statements is/are correct?

A I only
B II only
C I and II only
D I and III only
E II and III only

20.

The line spectrum for an element is shown below.

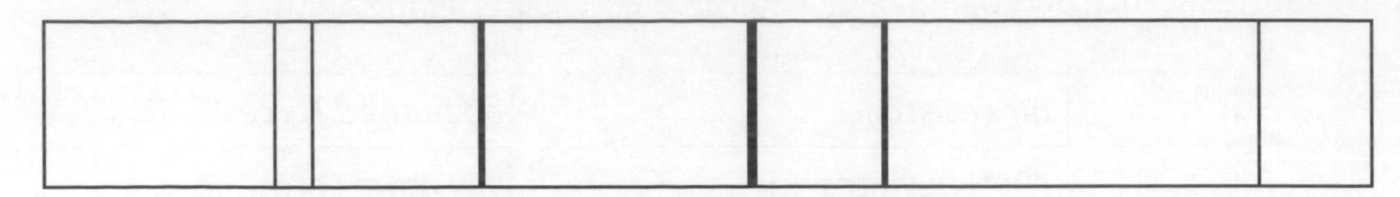

Line spectrum of element

The line spectra of different stars are shown below.

Identify which star this element is present in.

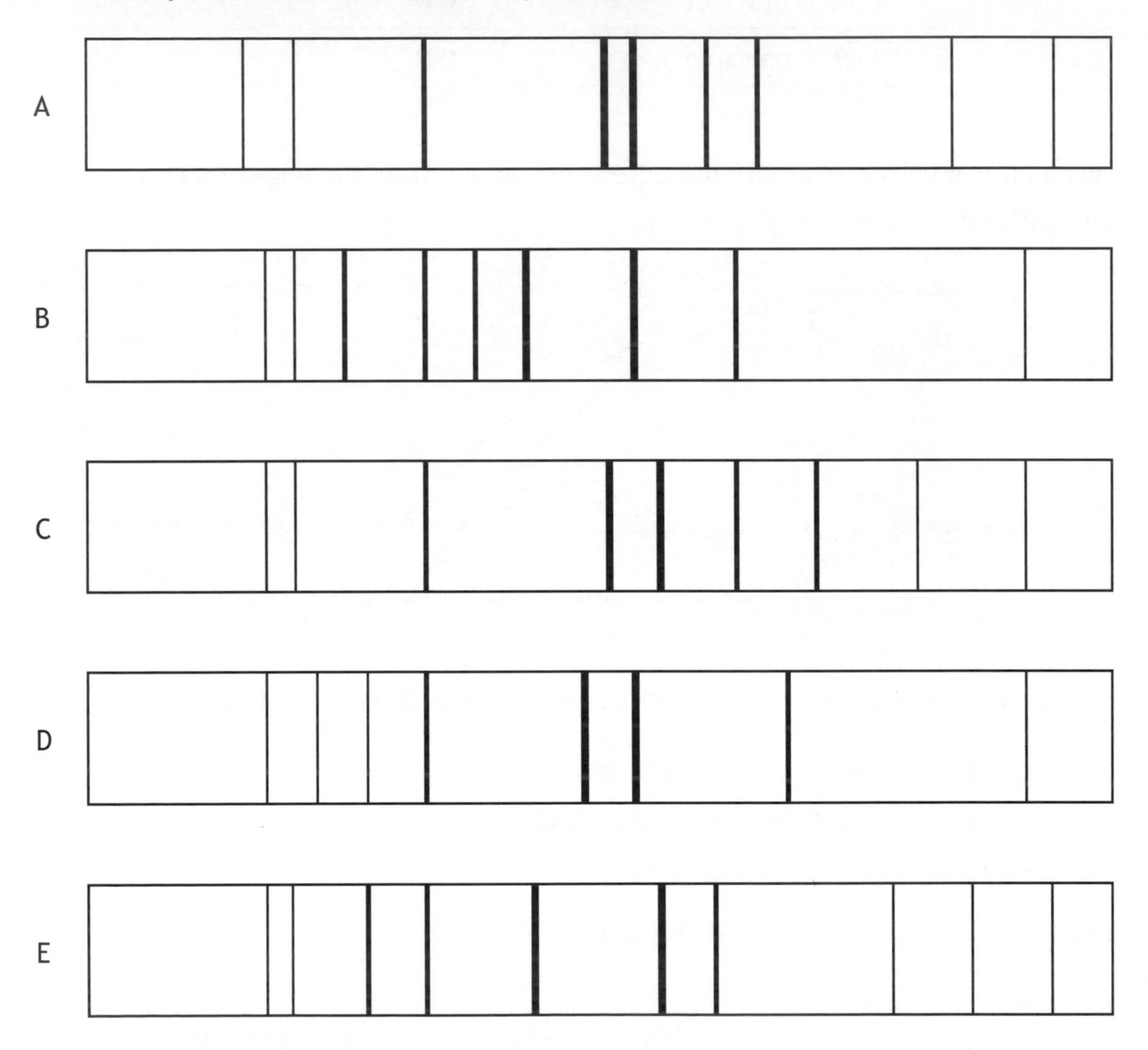

[END OF SECTION 1. NOW ATTEMPT THE QUESTIONS IN SECTION 2 OF YOUR QUESTION AND ANSWER BOOKLET]

National
Qualifications
MODEL PAPER 2

Physics
Relationships Sheet

Date — Not applicable

$E_p = mgh$

$E_k = \frac{1}{2}mv^2$

$Q = It$

$V = IR$

$R_T = R_1 + R_2 + ...$

$\frac{1}{R_T} = \frac{1}{R_1} + \frac{1}{R_2} + ...$

$V_2 = \left(\frac{R_2}{R_1 + R_2}\right)V_s$

$\frac{V_1}{V_2} = \frac{R_1}{R_2}$

$P = \frac{E}{t}$

$P = IV$

$P = I^2R$

$P = \frac{V^2}{R}$

$E_h = cm\Delta T$

$p = \frac{F}{A}$

$\frac{pV}{T} = \text{constant}$

$p_1V_1 = p_2V_2$

$\frac{p_1}{T_1} = \frac{p_2}{T_2}$

$\frac{V_1}{T_1} = \frac{V_2}{T_2}$

$d = vt$

$v = f\lambda$

$T = \frac{1}{f}$

$A = \frac{N}{t}$

$D = \frac{E}{m}$

$H = Dw_R$

$\dot{H} = \frac{H}{t}$

$s = vt$

$d = \bar{v}t$

$s = \bar{v}t$

$a = \frac{v-u}{t}$

$W = mg$

$F = ma$

$E_w = Fd$

$E_h = ml$

[END OF SPECIMEN RELATIONSHIPS SHEET]

National
Qualifications
MODEL PAPER 2

Physics Section 1— Answer Grid and Section 2

Duration — 2 hours

Total marks — 110

SECTION 1 — 20 marks

Attempt ALL questions in this section.

Instructions for completion of Section 1 are given on Page two.

SECTION 2 — 90 marks

Attempt ALL questions in this section.

Read all questions carefully before answering.

Use **blue** or **black** ink. Do NOT use gel pens.

Write your answers in the spaces provided. Additional space for answers and rough work is provided at the end of this booklet. If you use this space, write clearly the number of the question you are answering. Any rough work must be written in this booklet. You should score through your rough work when you have written your fair copy.

Before leaving the examination room you must give this booklet to the Invigilator. If you do not, you may lose all the marks for this paper.

SECTION 1 — 20 marks

The questions for Section 1 are contained in the booklet Physics Section 1 — Questions. Read these and record your answers on the grid on Page three opposite.

1. The answer to each question is **either** A, B, C, D or E. Decide what your answer is, then fill in the appropriate bubble (see sample question below).

2. There is **only one correct** answer to each question.

3. Any rough working should be done on the rough working sheet.

Sample Question

The energy unit measured by the electricity meter in your home is the:

A ampere

B kilowatt-hour

C watt

D coulomb

E volt.

The correct answer is **B**—kilowatt-hour. The answer **B** bubble has been clearly filled in (see below).

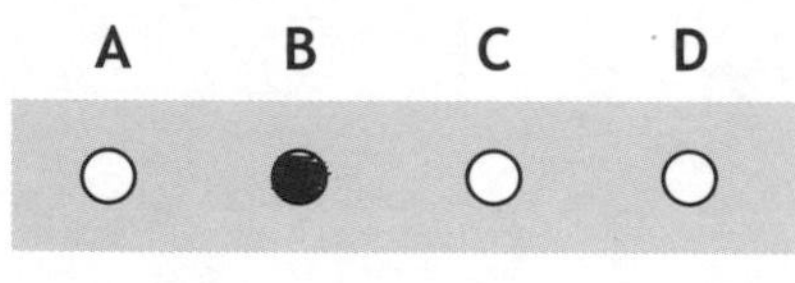

Changing an answer

If you decide to change your answer, cancel your first answer by putting a cross through it (see below) and fill in the answer you want. The answer below has been changed to **D**.

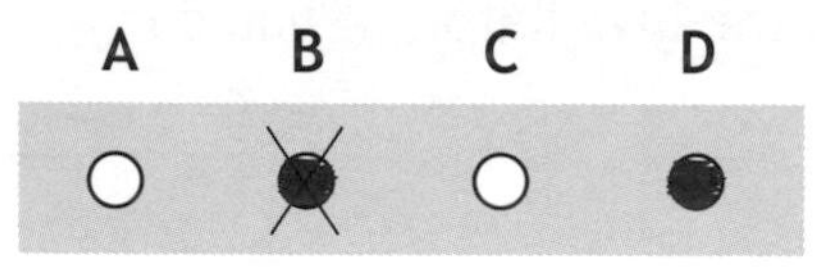

If you then decide to change back to an answer you have already scored out, put a tick (✓) to the **right** of the answer you want, as shown below:

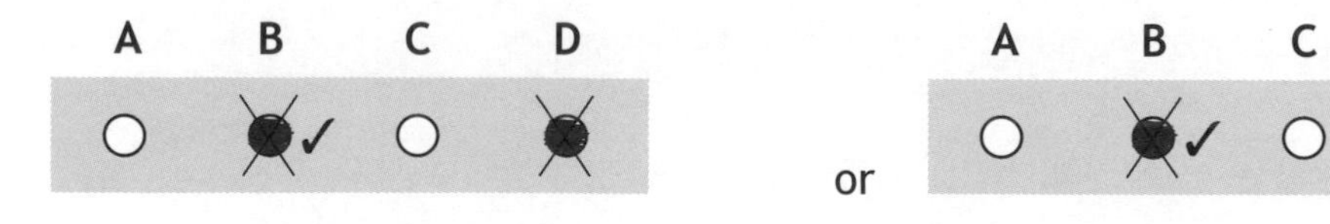

SECTION 1 — Answer Grid

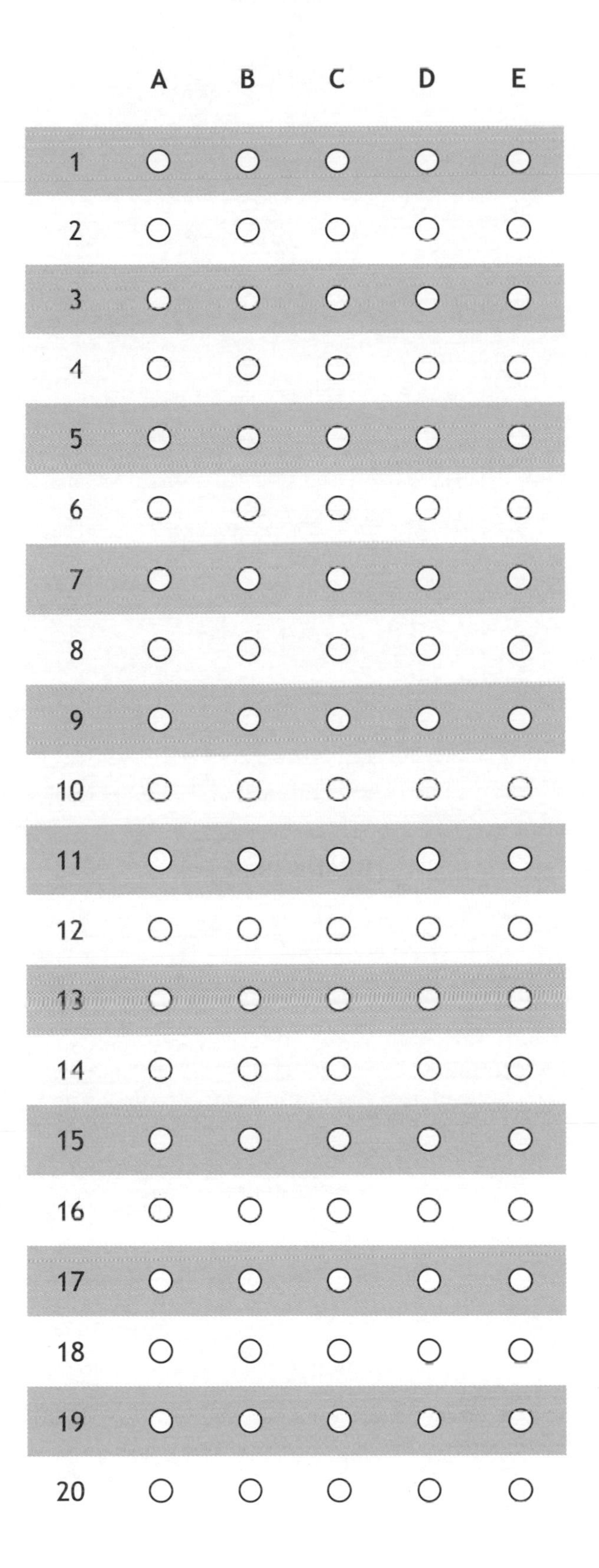

[BLANK PAGE]

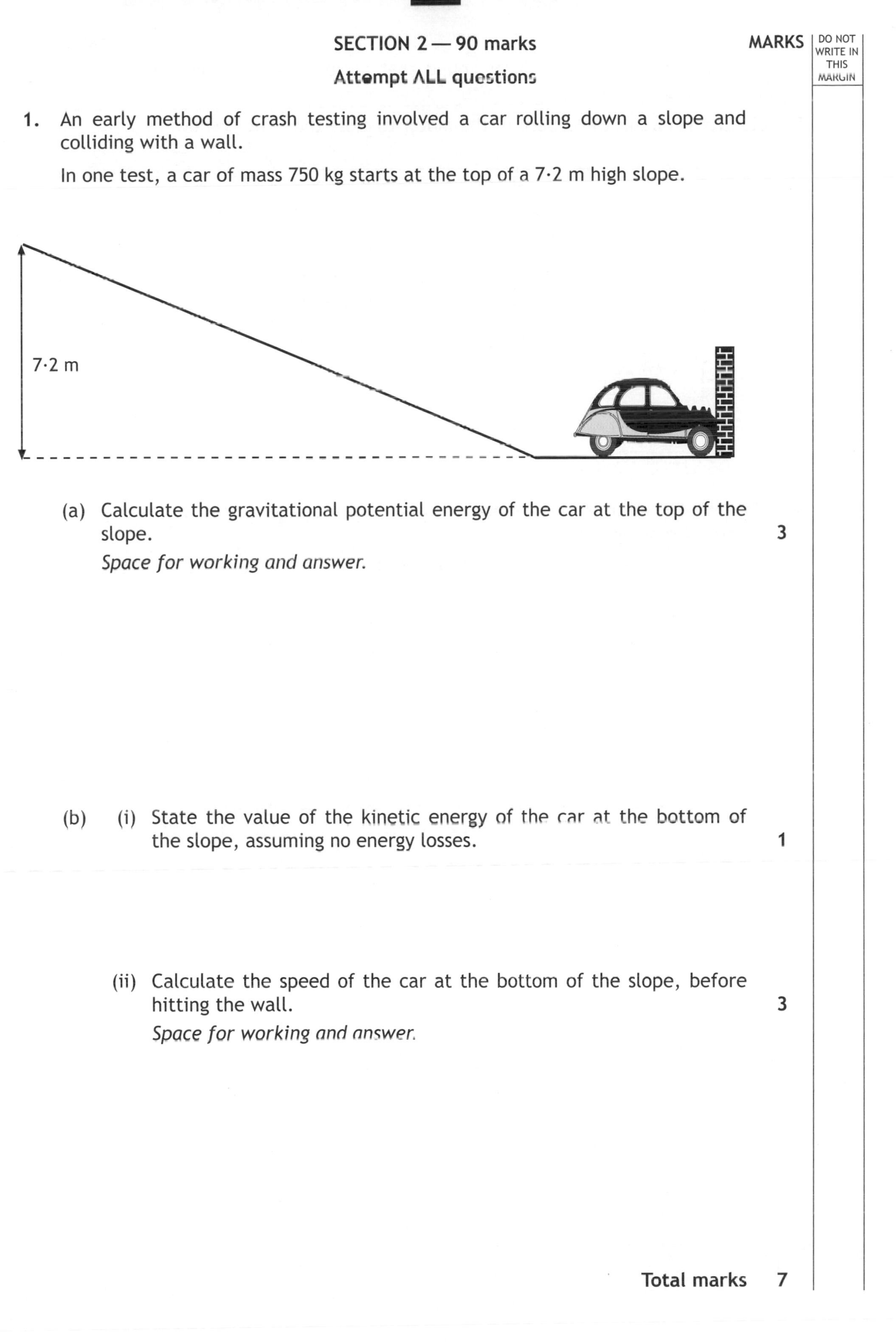

SECTION 2 — 90 marks

Attempt ALL questions

MARKS | DO NOT WRITE IN THIS MARGIN

1. An early method of crash testing involved a car rolling down a slope and colliding with a wall.

   In one test, a car of mass 750 kg starts at the top of a 7·2 m high slope.

   (a) Calculate the gravitational potential energy of the car at the top of the slope. **3**

   Space for working and answer.

   (b) (i) State the value of the kinetic energy of the car at the bottom of the slope, assuming no energy losses. **1**

   (ii) Calculate the speed of the car at the bottom of the slope, before hitting the wall. **3**

   Space for working and answer.

Total marks 7

MARKS DO NOT WRITE IN THIS MARGIN

2. An office has an automatic window blind that closes when the light level outside gets too high.

The electronic circuit that operates the motor to close the blind is shown.

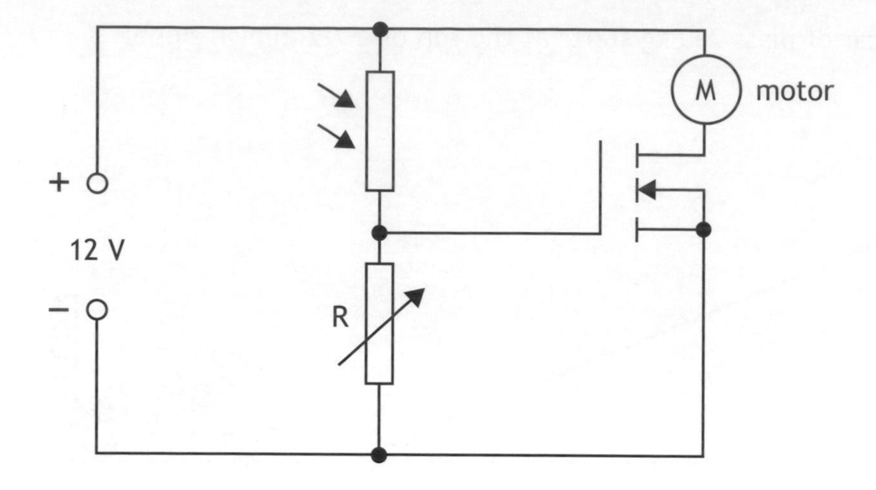

(a) The MOSFET transistor switches on when the voltage across variable resistor R reaches 2·4 V.

Explain how this circuit works to close the blind. **2**

(b) The graph shows how the resistance of the LDR varies with light level.

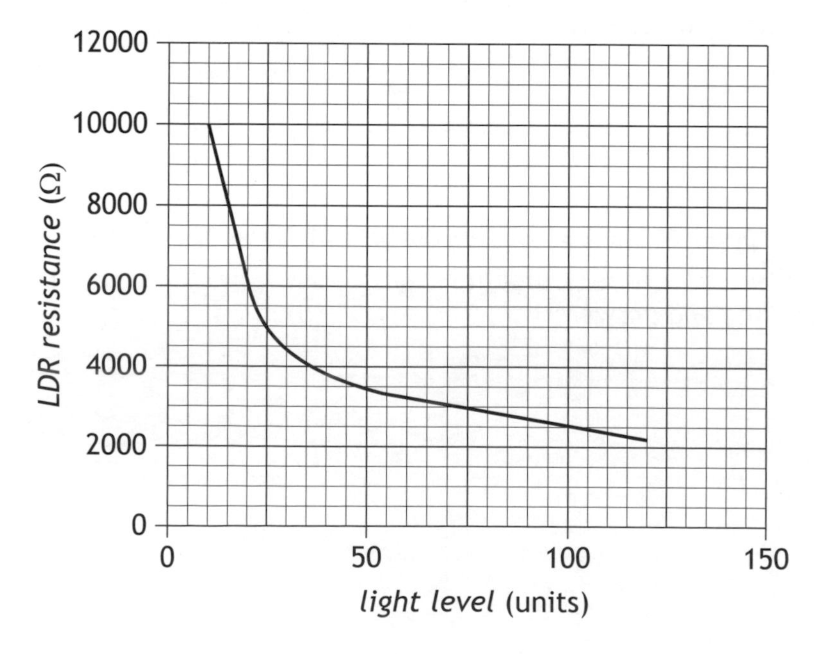

MARKS DO NOT WRITE IN THIS MARGIN

2. (continued)

(i) What is the resistance of the LDR when the light level is 70 units? **1**

(ii) R has a value of 600 Ω.

Calculate the voltage across R when the light level is 70 units. **3**

Space for working and answer.

(iii) State whether or not the blinds will close when the light level is 70 units.

You must justify your answer. **2**

Total marks 9

MARKS | DO NOT WRITE IN THIS MARGIN

3. (a) State the definition of pressure. **1**

(b) To remain safe when diving, deep sea divers must know the pressure exerted on them by the sea at different depths.

The pressure exerted on deep sea divers when diving beneath the sea is calculated using the relationship:

$$p = \rho gh$$

where:

p is the pressure exerted by the sea in Pa

ρ is the density of the sea water in kg m^{-3}

g is the acceleration due to gravity in m s^{-2}

h is the submerged depth of the diver in metres.

During one dive, a diver reaches a depth of 24 m. The density of the water is 1025 kg m^{-3}.

Calculate the pressure exerted by the sea water on the diver at this depth. **3**

Space for working and answer.

Total marks 4

MARKS DO NOT WRITE IN THIS MARGIN

4. Car designers are constantly trying to reduce the environmental impact of cars. One way to do this is to make them more fuel-efficient, as the less fuel cars need, the fewer dangerous gases they emit into the atmosphere.

Use your knowledge of physics to comment on how car manufacturers might produce cars which are more fuel efficient. **3**

MARKS DO NOT WRITE IN THIS MARGIN

5. A student carries out an experiment to investigate the relationship between the pressure and temperature of a fixed mass of gas. The apparatus used is shown.

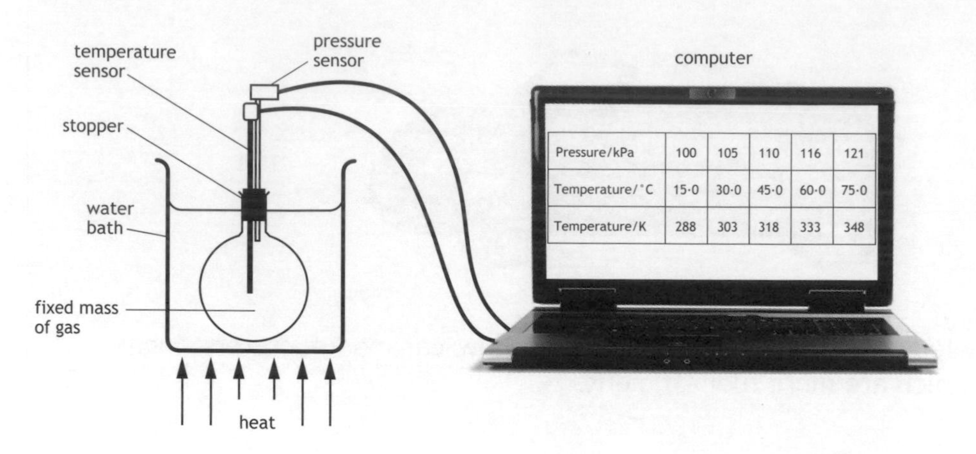

The pressure and temperature of the gas are recorded using sensors connected to a computer. The gas is heated slowly in the water bath and a series of readings is taken.

The volume of the gas remains constant during the experiment.

The results are shown.

Pressure/kPa	100	105	110	116	121
Temperature/°C	15·0	30·0	45·0	60·0	75·0
Temperature/K	288	303	318	333	348

MARKS | DO NOT WRITE IN THIS MARGIN

5. **(continued)**

(a) Using **all** the relevant data, establish the relationship between the pressure and the temperature of the gas. **2**

(b) Use the kinetic model to explain the change in pressure as the temperature of the gas increases. **3**

(c) Explain why the level of water in the water bath should be above the bottom of the stopper. **1**

Total marks 6

MARKS | DO NOT WRITE IN THIS MARGIN

6. A satellite sends microwaves to a ground station on Earth.

(a) The microwaves have a wavelength of 60 mm.

(i) Calculate the frequency of the waves. **3**

Space for working and answer.

(ii) Calculate the period of the waves. **3**

Space for working and answer.

(b) The satellite sends radio waves along with the microwaves to the ground station. Will the radio waves be received by the ground station **before**, **after** or **at the same time** as the microwaves?

Explain your answer. **2**

MARKS DO NOT WRITE IN THIS MARGIN

6. (continued)

(c) A music concert is being broadcast live on radio.

Drivers in two cars, A and B, are listening to the performance on the radio.

The performance is being broadcast on two different wavebands, from the same transmitter.

The radio in car A is tuned to a radio signal of frequency 1152 kHz.

The radio in car B is tuned to a radio signal of frequency 102·5 MHz.

Both cars drive into a valley surrounded by hills.

The radio in car B loses the signal from the broadcast.

Explain why this signal is lost. **2**

Total marks 10

MARKS DO NOT WRITE IN THIS MARGIN

7. Gold-198 is a radioactive source that is used to trace factory waste which may cause river pollution.

A small quantity of the radioactive gold is added into the waste as it enters the river. Scanning the river using radiation detectors allows scientists to trace where the waste has travelled.

Gold-198 has a half-life of 2·7 days.

(a) What is meant by the term "half-life"? **1**

(b) A sample of Gold-198 has an activity of 64 kBq when first obtained by the scientists.

Calculate the activity after 13·5 days. **3**

Space for working and answer.

(c) Describe two precautions taken by the scientists to reduce the equivalent dose they receive while using radioactive sources. **2**

Total marks 6

MARKS DO NOT WRITE IN THIS MARGIN

8. Many countries use nuclear reactors to produce energy.

Nuclear fission occurs inside the reactor.

A diagram of the core of a nuclear reactor is shown.

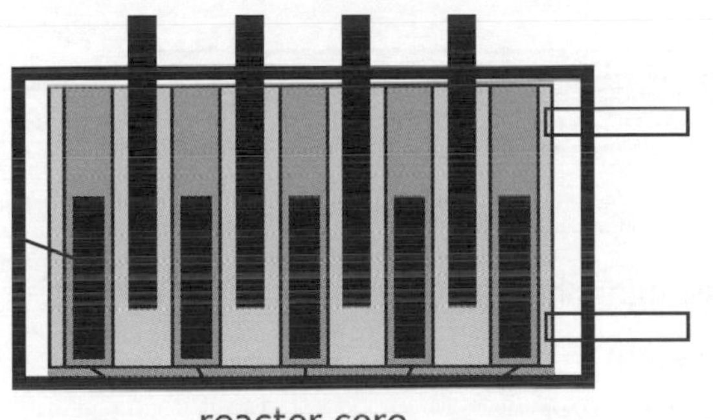

reactor core

(a) Describe what happens when a nuclear fission reaction occurs. **2**

(b) One nuclear fission reaction produces $2{\cdot}9 \times 10^{-11}$ J of energy.

The power output of the reactor is 1·4 GW.

How many fission reactions are produced in one hour? **4**

Space for working and answer.

(c) State **one disadvantage** of using nuclear power for the generation of electricity. **1**

Total marks 7

MARKS | DO NOT WRITE IN THIS MARGIN

9. An aircraft is flying horizontally at a constant speed.

(a) The aircraft and passengers have a total mass of 50 000 kg.

Calculate the total weight. **3**

Space for working and answer.

(b) State the magnitude of the upward force acting on the aircraft. **1**

MARKS DO NOT WRITE IN THIS MARGIN

9. (continued)

(c) During the flight, the aircraft's engines produce a force of $4{\cdot}4 \times 10^4$ N due North. The aircraft encounters a crosswind, blowing from west to east, which exerts a force of $3{\cdot}2 \times 10^4$ N.

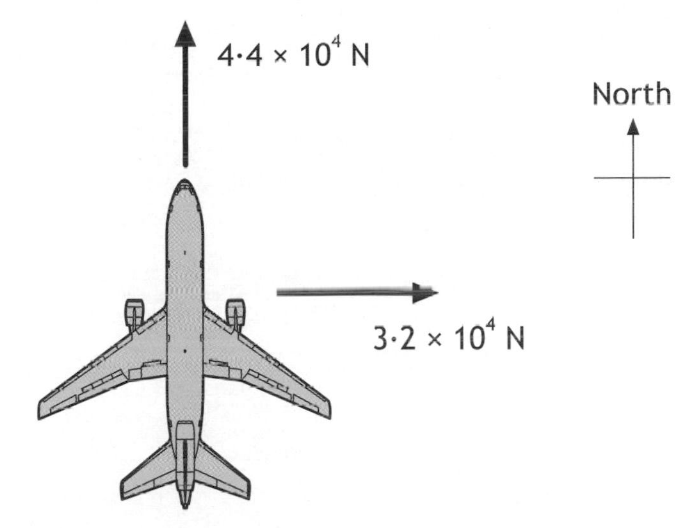

Calculate the resultant force on the aircraft. **4**

Space for working and answer.

MARKS | DO NOT WRITE IN THIS MARGIN

9. (continued)

(d) During a particular flight, a pilot receives an absorbed dose of 15 μGy from gamma rays.

Calculate the equivalent dose received due to this type of radiation. **4**

Space for working and answer.

(e) Gamma radiation is an example of radiation which causes ionisation.

Explain what is meant by the term ionisation. **1**

Total marks 12

MARKS DO NOT WRITE IN THIS MARGIN

10. Athletes in a race are recorded by a TV camera which runs on rails beside the track.

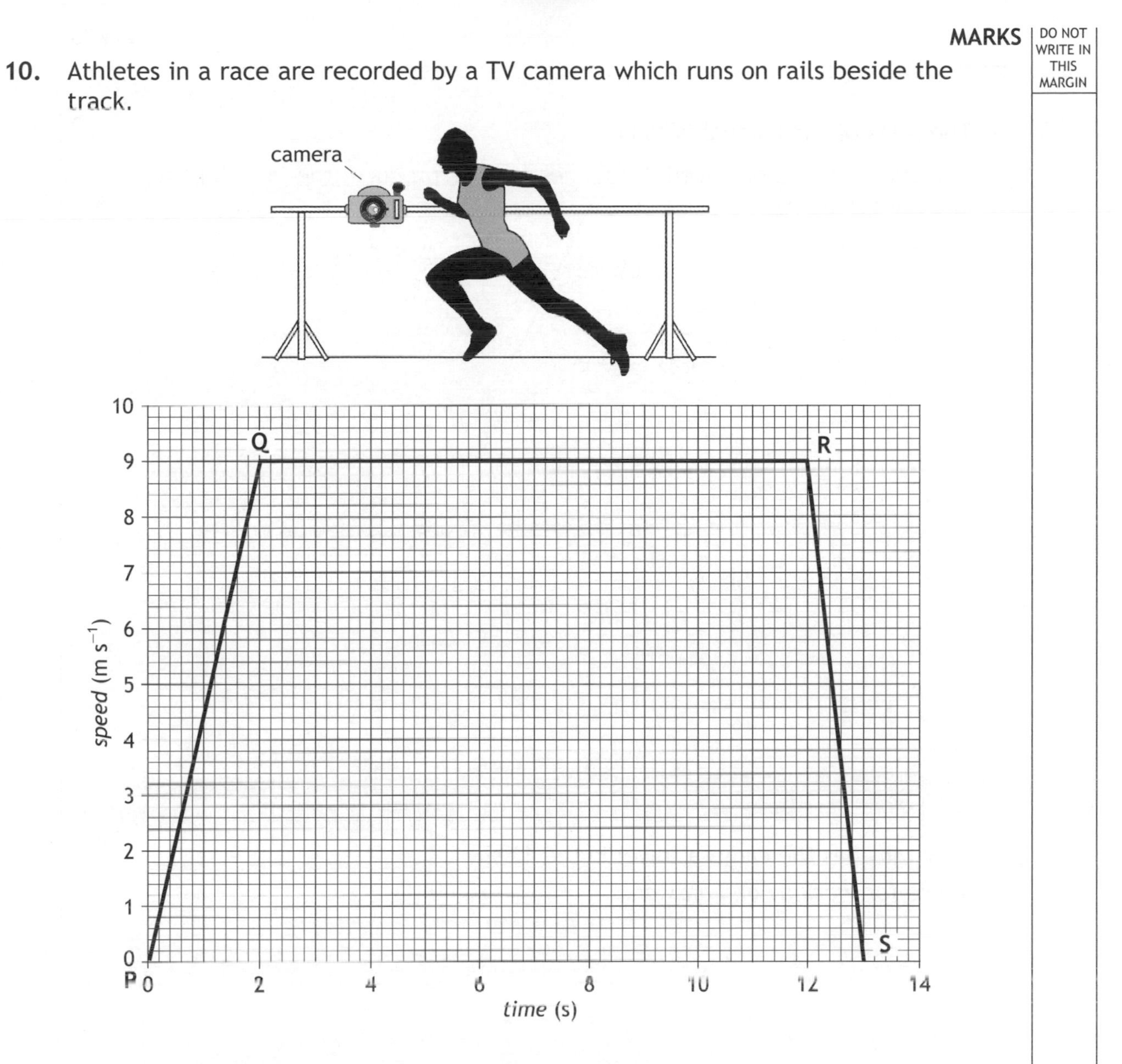

The graph shows the speed of the camera during the race.

(a) Calculate the acceleration of the camera between **P** and **Q**. **3**

Space for working and answer.

MARKS | DO NOT WRITE IN THIS MARGIN

10. (continued)

(b) The mass of the camera is 15 kg.

Calculate the unbalanced force needed to produce the acceleration between **P** and **Q**. **3**

Space for working and answer.

(c) How far does the camera travel in the 13 s? **3**

Space for working and answer.

Total marks 9

MARKS DO NOT WRITE IN THIS MARGIN

11. One type of exercise machine is shown below.

(a) A person using this machine pedals against friction forces applied to the wheel by the brake.

A friction force of 300 N is applied at the edge of the wheel, which has a circumference of 1.5 m.

How much work is done by friction in 500 turns of the wheel? **4**

Space for working and answer.

MARKS DO NOT WRITE IN THIS MARGIN

11. (continued)

(b) The wheel is a solid aluminium disc of mass 12.0 kg.

(i) All the work done by friction is converted to heat in the disc.

Calculate the temperature rise after 500 turns. 4

Space for working and answer.

(ii) Explain why the actual temperature rise of the disc is less than calculated in (b)(i) 1

Total marks 9

MARKS DO NOT WRITE IN THIS MARGIN

12. When a spacecraft is launched into space it accelerates to reach speeds of up to 8 km s^{-1} to achieve orbit.

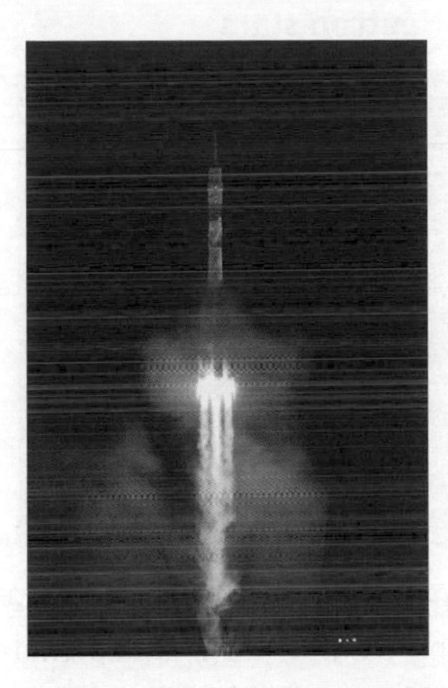

At launch, most of its mass consists of the fuel required to provide upthrust for this acceleration.

During the launch, the acceleration of the spacecraft is not constant.

Use your knowledge of physics to comment on why the acceleration is not constant. **3**

MARKS | DO NOT WRITE IN THIS MARGIN

13. Read the passage below and answer the questions that follow.

Neutron stars

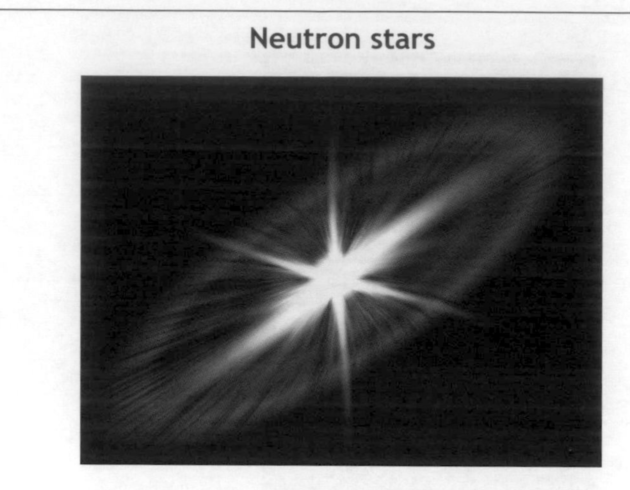

When stars reach the end of their life, they can become neutron stars. A neutron star has a mass ranging from 1·4-3·2 times that of our Sun. In a neutron star, this huge mass is contained within a diameter of approximately 12 km, and means that a neutron star is extremely dense.

One cupful of this mass would have the same weight as Mount Everest on Earth!

The extreme density of the neutron star also means that it has very large gravitational and magnetic field strengths.

Stars emit huge amounts of energy. This energy is the result of nuclear fusion happening at the centre of the star. Nuclear fusion of the isotopes of hydrogen produces helium and also the energy which sustains the star's massive shape.

Neutron stars are thought to be formed when large stars collapse.

This happens when the fusion process stops and there is no longer enough energy to sustain the star. The star explodes.

This explosion is known as a supernova. The outer gases of the star expand rapidly to produce an extremely bright object in the sky, which can be seen by astronomers on Earth.

The gravitational field causes the centre of the star to collapse. Its volume reduces dramatically. During the collapsing process, electrons and protons combine to form neutrons. This is the reason for the name "neutron" stars.

Neutron stars sometimes appear in binary systems, where they are in mutual orbit around another object. X-ray telescopes on satellites have been used by astronomers to obtain data from such binary systems. This data has confirmed the mass of the neutron star to be 1·4–3·2 times that of the Sun's mass.

Neutron stars rotate rapidly when newly formed, and gradually slow over a long period of time. A neutron star, known as PSR J1748-2446ad, rotates 716 times per second.

Some neutron stars emit radio waves or X-rays. These emissions only occur at the magnetic poles of the neutron star. When observed by astronomers, these emissions appear as "pulses" of radio waves or X-rays. The pulses appear at the same rate as the rotation of the neutron star. Such neutron stars are known as "pulsars".

MARKS | DO NOT WRITE IN THIS MARGIN

13. (continued)

(a) What event is thought to lead to the formation of neutron stars? **1**

(b) Why does the neutron star consist mainly of neutrons? **1**

(c) Calculate the period of rotation of the neutron star known as PSR J1748-2446ad. **3**

Space for working and answer.

Total marks 5

[END OF MODEL PAPER]

MARKS DO NOT WRITE IN THIS MARGIN

ADDITIONAL SPACE FOR ROUGH WORKING AND ANSWERS

MARKS DO NOT WRITE IN THIS MARGIN

ADDITIONAL SPACE FOR ROUGH WORKING AND ANSWERS

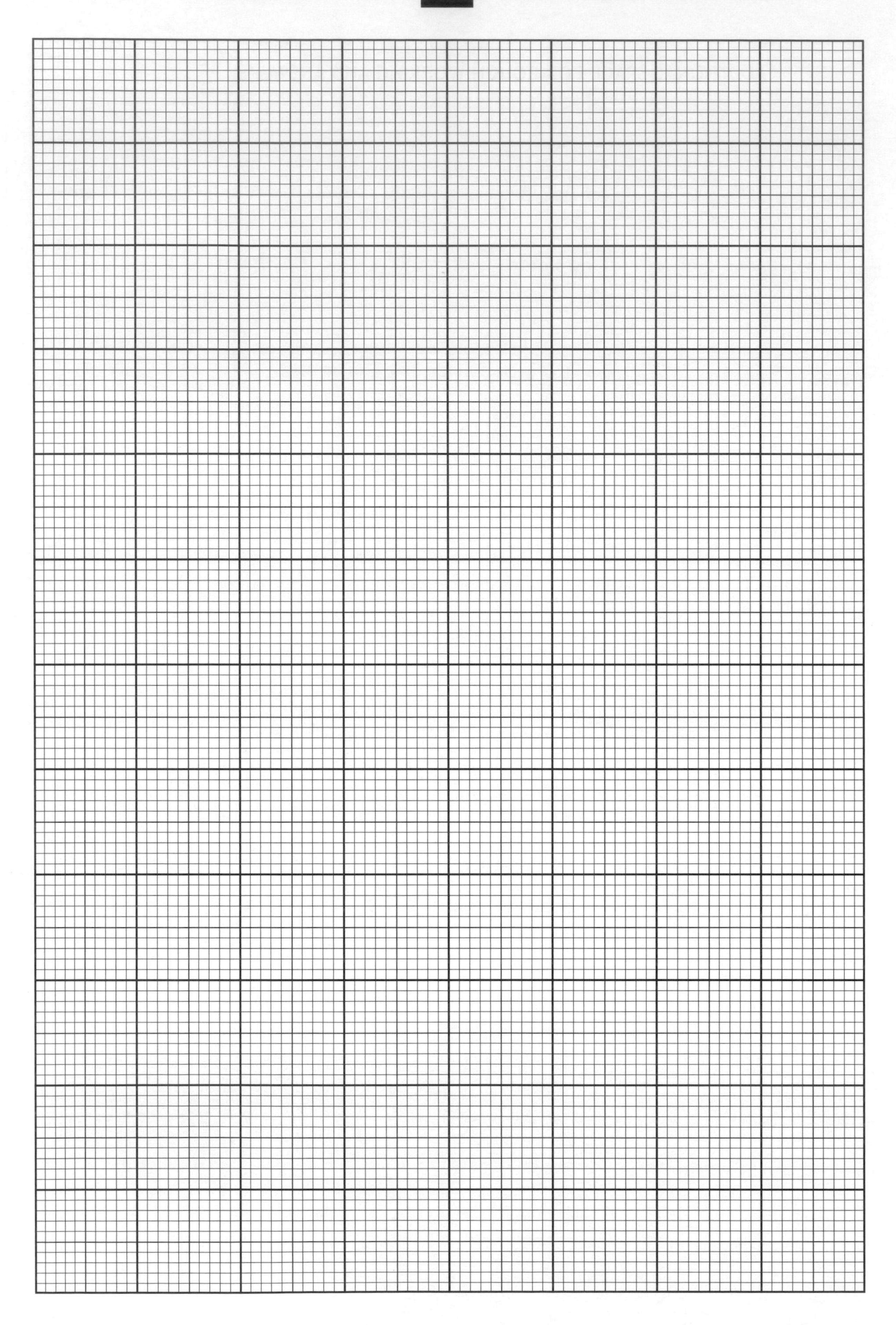

NATIONAL 5

Model Paper 3

Whilst this Model Paper has been specially commissioned by Hodder Gibson for use as practice for the National 5 exams, the key reference documents remain the SQA Specimen Paper 2013 and the SQA Past Papers 2014 and 2015.

National
Qualifications
MODEL PAPER 3

Physics
Section 1—Questions

Date — Not applicable

Duration — 2 hours

Instructions for completion of Section 1 are given on Page two of the question paper.

Record your answers on the grid on Page three of your answer booklet

Do NOT write in this booklet.

Before leaving the examination room you must give your answer booklet to the Invigilator. If you do not, you may lose ALL the marks for this paper.

DATA SHEET

Speed of light in materials

Material	*Speed* in $m\,s^{-1}$
Air	$3{\cdot}0 \times 10^8$
Carbon dioxide	$3{\cdot}0 \times 10^8$
Diamond	$1{\cdot}2 \times 10^8$
Glass	$2{\cdot}0 \times 10^8$
Glycerol	$2{\cdot}1 \times 10^8$
Water	$2{\cdot}3 \times 10^8$

Gravitational field strengths

	Gravitational field strength on the surface in $N\,kg^{-1}$
Earth	9·8
Jupiter	23
Mars	3·7
Mercury	3·7
Moon	1·6
Neptune	11
Saturn	9·0
Sun	270
Uranus	8·7
Venus	8·9

Specific latent heat of fusion of materials

Material	*Specific latent heat of fusion* in $J\,kg^{-1}$
Alcohol	$0{\cdot}99 \times 10^5$
Aluminium	$3{\cdot}95 \times 10^5$
Carbon dioxide	$1{\cdot}80 \times 10^5$
Copper	$2{\cdot}05 \times 10^5$
Iron	$2{\cdot}67 \times 10^5$
Lead	$0{\cdot}25 \times 10^5$
Water	$3{\cdot}34 \times 10^5$

Specific latent heat of vaporisation of materials

Material	*Specific latent heat of vaporisation* in $J\,kg^{-1}$
Alcohol	$11{\cdot}2 \times 10^5$
Carbon dioxide	$3{\cdot}77 \times 10^5$
Glycerol	$8{\cdot}30 \times 10^5$
Turpentine	$2{\cdot}90 \times 10^5$
Water	$22{\cdot}6 \times 10^5$

Speed of sound in materials

Material	*Speed* in $m\,s^{-1}$
Aluminium	5200
Air	340
Bone	4100
Carbon dioxide	270
Glycerol	1900
Muscle	1600
Steel	5200
Tissue	1500
Water	1500

Specific heat capacity of materials

Material	*Specific heat capacity* in $J\,kg^{-1}\,°C^{-1}$
Alcohol	2350
Aluminium	902
Copper	386
Glass	500
Ice	2100
Iron	480
Lead	128
Oil	2130
Water	4180

Melting and boiling points of materials

Material	*Melting point* in °C	*Boiling point* in °C
Alcohol	−98	65
Aluminium	660	2470
Copper	1077	2567
Glycerol	18	290
Lead	328	1737
Iron	1537	2737

Radiation weighting factors

Type of radiation	*Radiation weighting factor*
alpha	20
beta	1
fast neutrons	10
gamma	1
slow neutrons	3

SECTION 1

1. The unit of current is the ampere.

One ampere can also be expressed as

A one volt per joule

B one coulomb per second

C one ohm per volt

D one joule per second

E one joule per coulomb.

2. The potential difference across a lamp is 8 V when the current is 2 A.

The charge which passes through the lamp in three minutes is

A 6 C

B 16 C

C 18 C

D 360 C

E 1080 C.

3. Which diagram correctly shows the direction which atomic particles electrons, protons and neutrons are deflected by an electric field?

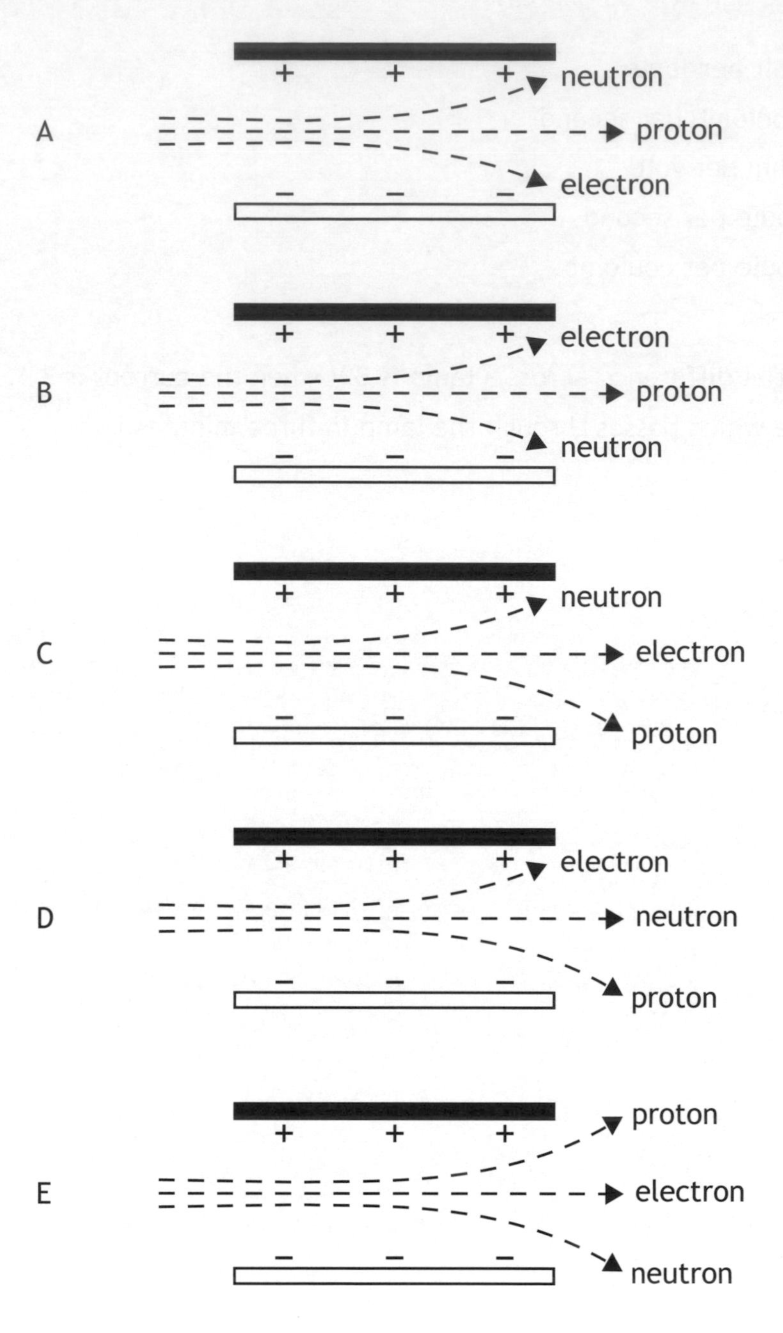

4. The graph below shows how the current is related to the applied potential difference for two separate resistors P and Q.

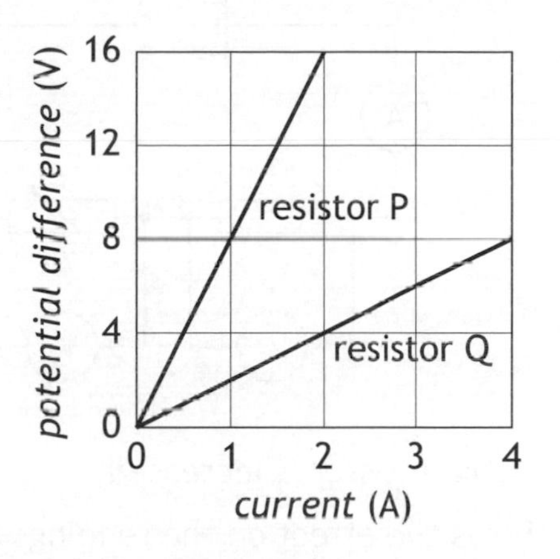

The values of the two resistors shown, in ohms, are

	Resistance of P (Ω)	*Resistance of Q* (Ω)
A	0·125	0·5
B	2	4
C	8	2
D	8	16
E	16	8

5. A circuit is set up as shown.

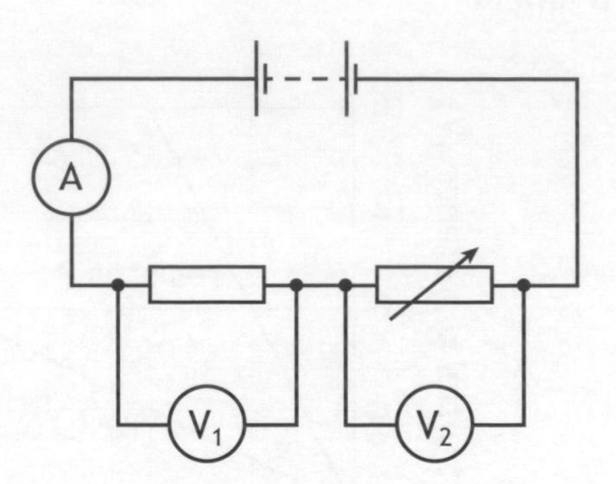

The resistance of the variable resistor is increased.

Which row in the table shows the effect on the readings on the ammeter and voltmeters?

	Reading on ammeter	*Reading on voltmeter V_1*	*Reading on voltmeter V_2*
A	decreases	decreases	decreases
B	increases	unchanged	increases
C	decreases	increases	decreases
D	increases	unchanged	decreases
E	decreases	decreases	increases

6. A circuit is set up as shown.

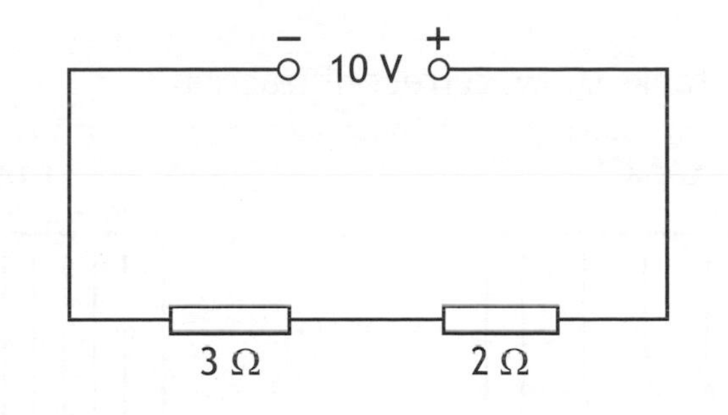

The potential difference across the 2 Ω resistor is

A 4 V

B 5 V

C 6 V

D 10 V

E 20 V.

7. Which of the following describes the frequency of a water wave?

A The distance between the crest of a wave and the crest of the next wave

B The time taken for one complete wave to pass any point

C The number of waves passing any point in one second

D The distance travelled by a crest in one second

E The time taken for the source to make one complete vibration

8. Water waves are produced in two identical ripple tanks. The waves reach a barrier and are diffracted.

Which pair of ripple tanks shows correct diffraction?

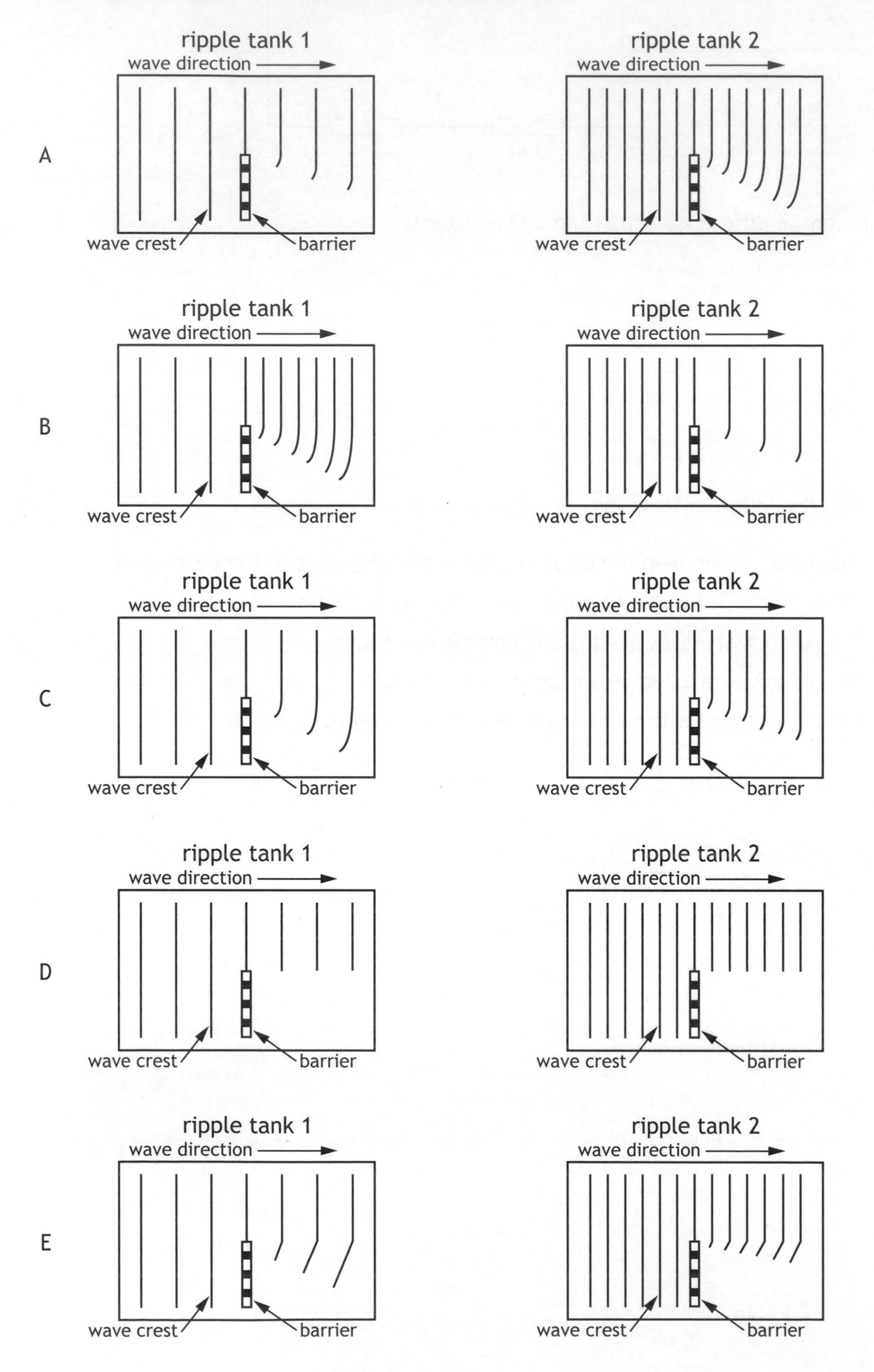

9. Which of the following electromagnetic waves has a higher frequency than visible light and a lower frequency than X-rays?

A Gamma rays

B Infrared

C Microwaves

D Radio

E Ultraviolet

10. Which of the following describes the term ionisation?

A An atom losing an orbiting electron

B An atom losing a proton

C A nucleus emitting an alpha particle

D A nucleus emitting a neutron

E A nucleus emitting a gamma ray

11. Compared with a proton, an alpha particle has

A twice the mass and twice the charge

B twice the mass and the same charge

C four times the mass and twice the charge

D four times the mass and the same charge

E twice the mass and four times the charge.

12. One gray is equal to

A one becquerel per kilogram

B one sievert per second

C one joule per second

D one sievert per kilogram

E one joule per kilogram.

13. A sample of a radioactive material has a mass of 30 g. There are 36 000 nuclear decays every minute in this sample.

The activity of the sample is

A 600 Bq

B 1800 Bq

C 36 000 Bq

D 1 080 000 Bq

E 2 160 000 Bq.

14. Which of the following statements about nuclear fusion is correct?

A Energy is released when a nucleus with a large mass number splits into two nuclei of smaller mass number.

B Energy is absorbed when a nucleus with a large mass number splits into two nuclei of smaller mass number.

C Energy is absorbed when two nuclei combine to form a nucleus of larger mass number.

D Energy is released when two nuclei combine to form a nucleus of larger mass number.

E Energy is absorbed when a nucleus with a large mass number combines with a nucleus of small mass number to produce a nucleus of larger mass number.

15. Which row contains two scalar quantities and one vector quantity?

A Distance, weight, velocity

B Speed, mass, displacement

C Distance, weight, force

D Speed, weight, acceleration

E Velocity, force, mass

16. A student follows the route shown in the diagram and arrives back at the starting point.

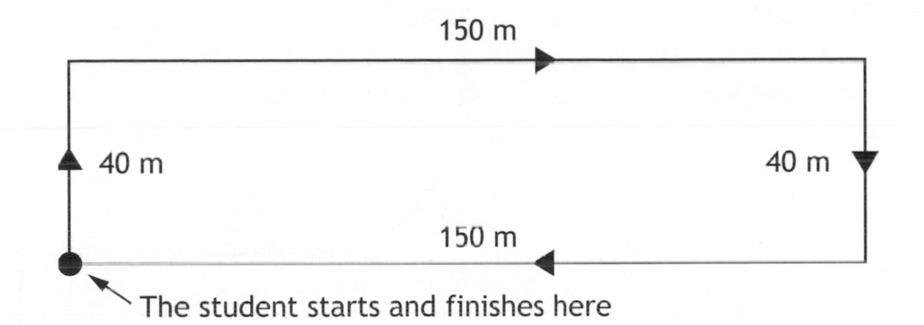

Which row in the table shows the total distance walked and the magnitude of the final displacement?

	Total distance (m)	*Final displacement* (m)
A	0	80
B	0	380
C	190	0
D	380	0
E	380	380

17. The graph shows how the velocity of an object varies with time.

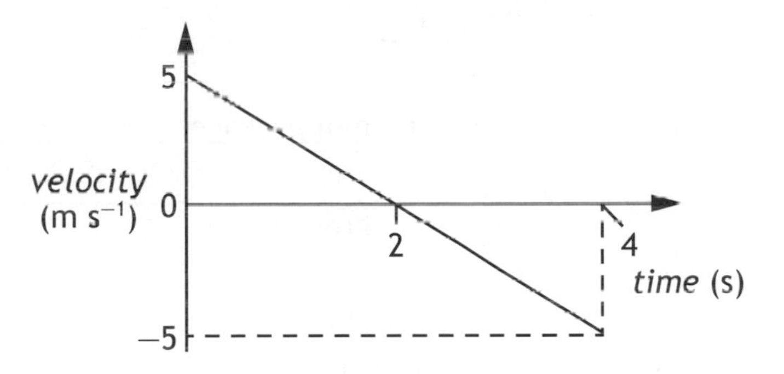

Which row in the table shows the displacement after 4 s and the acceleration of the object during the first 4 s?

	Displacement (m)	*Acceleration* ($m\ s^{-2}$)
A	10	−10
B	10	2·5
C	0	2·5
D	0	−10
E	0	−2·5

18. A ball was dropped into a deep well. The graph shows the speed of the ball from the instant of its release in air until it has fallen several metres through the water to the bottom of the well.

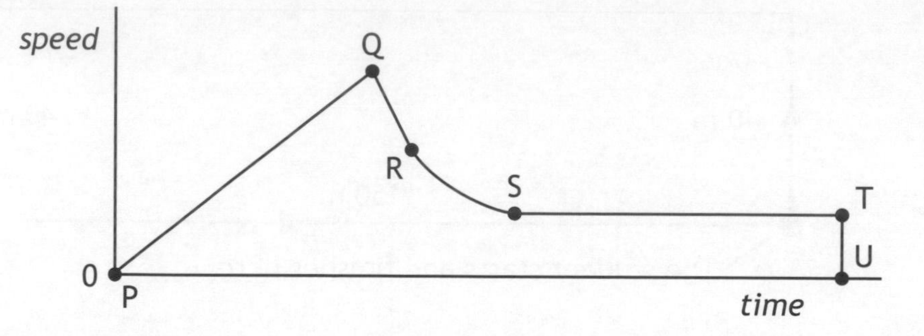

Which part of the graph indicates when the vertical forces acting on the ball were balanced?

A PQ

B QR

C RS

D ST

E TU

19. A package is released from a helicopter flying horizontally at a constant speed of 39 m s^{-1}.

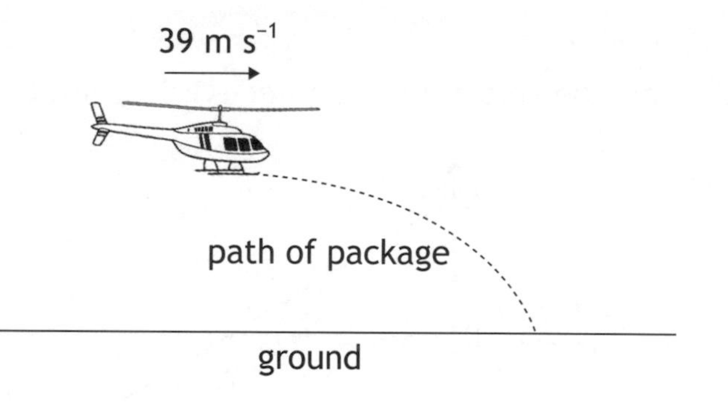

The package takes 5·0 s to reach the ground.

The effects of air resistance can be ignored.

Which row in the table shows the horizontal speed and vertical speed of the package just before it hits the ground?

	Horizontal speed (m s^{-1})	*Vertical speed* (m s^{-1})
A	0	39
B	39	39
C	39	49
D	49	39
E	49	49

20. The diameter of the disk of the Milky Way galaxy spans a distance of about 100 000 light years.

The diameter of the Milky Way galaxy, expressed in metres is

A $9{\cdot}5 \times 10^{5}$

B $3{\cdot}0 \times 10^{13}$

C $1{\cdot}1 \times 10^{16}$

D $2{\cdot}6 \times 10^{17}$

E $9{\cdot}5 \times 10^{20}$.

[END OF SECTION 1. NOW ATTEMPT THE QUESTIONS IN SECTION 2 OF YOUR QUESTION AND ANSWER BOOKLET]

[BLANK PAGE]

National
Qualifications
MODEL PAPER 3

Physics
Relationships Sheet

Date — Not applicable

$E_p = mgh$

$E_k = \frac{1}{2}mv^2$

$Q = It$

$V = IR$

$R_T = R_1 + R_2 + \ldots$

$\frac{1}{R_T} = \frac{1}{R_1} + \frac{1}{R_2} + \ldots$

$V_2 = \left(\frac{R_2}{R_1 + R_2}\right)V_s$

$\frac{V_1}{V_2} = \frac{R_1}{R_2}$

$P = \frac{E}{t}$

$P = IV$

$P = I^2R$

$P = \frac{V^2}{R}$

$E_h = cm\Delta T$

$p = \frac{F}{A}$

$\frac{pV}{T} = \text{constant}$

$p_1V_1 = p_2V_2$

$\frac{p_1}{T_1} = \frac{p_2}{T_2}$

$\frac{V_1}{T_1} = \frac{V_2}{T_2}$

$d = vt$

$v = f\lambda$

$T = \frac{1}{f}$

$A = \frac{N}{t}$

$D = \frac{E}{m}$

$H = Dw_R$

$\dot{H} = \frac{H}{t}$

$s = vt$

$d = \bar{v}t$

$s = \bar{v}t$

$a = \frac{v-u}{t}$

$W = mg$

$F = ma$

$E_w = Fd$

$E_h = ml$

[END OF SPECIMEN RELATIONSHIPS SHEET]

National
Qualifications
MODEL PAPER 3

Physics Section 1— Answer Grid and Section 2

Duration — 2 hours

Total marks — 110

SECTION 1 — 20 marks

Attempt ALL questions in this section.

Instructions for completion of Section 1 are given on Page two.

SECTION 2 — 90 marks

Attempt ALL questions in this section.

Read all questions carefully before answering.

Use **blue** or **black** ink. Do NOT use gel pens.

Write your answers in the spaces provided. Additional space for answers and rough work is provided at the end of this booklet. If you use this space, write clearly the number of the question you are answering. Any rough work must be written in this booklet. You should score through your rough work when you have written your fair copy.

Before leaving the examination room you must give this booklet to the Invigilator. If you do not, you may lose all the marks for this paper.

SECTION 1 — 20 marks

The questions for Section 1 are contained in the booklet Physics Section 1 — Questions.
Read these and record your answers on the grid on Page three opposite.

1. The answer to each question is **either** A, B, C, D or E. Decide what your answer is, then fill in the appropriate bubble (see sample question below).

2. There is **only one correct** answer to each question.

3. Any rough working should be done on the rough working sheet.

Sample Question

The energy unit measured by the electricity meter in your home is the:

A ampere

B kilowatt-hour

C watt

D coulomb

E volt.

The correct answer is **B**—kilowatt-hour. The answer **B** bubble has been clearly filled in (see below).

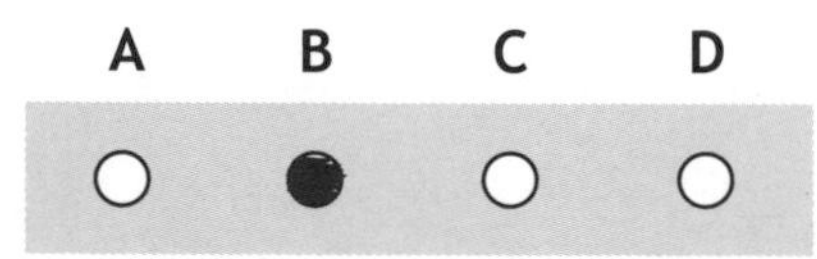

Changing an answer

If you decide to change your answer, cancel your first answer by putting a cross through it (see below) and fill in the answer you want. The answer below has been changed to **D**.

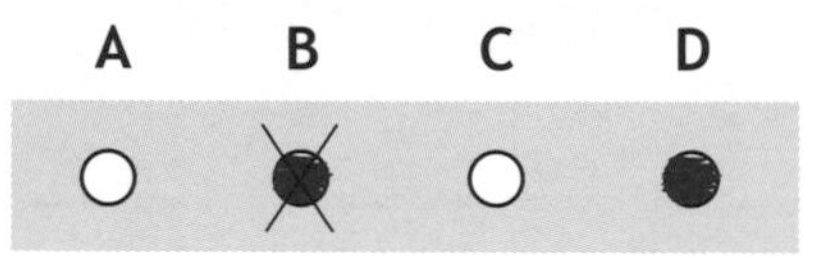

If you then decide to change back to an answer you have already scored out, put a tick (✓) to the **right** of the answer you want, as shown below:

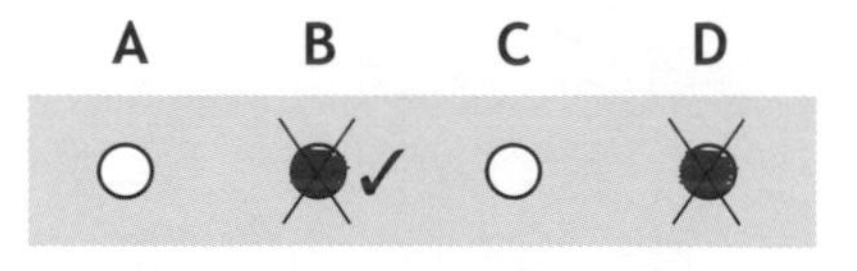

or

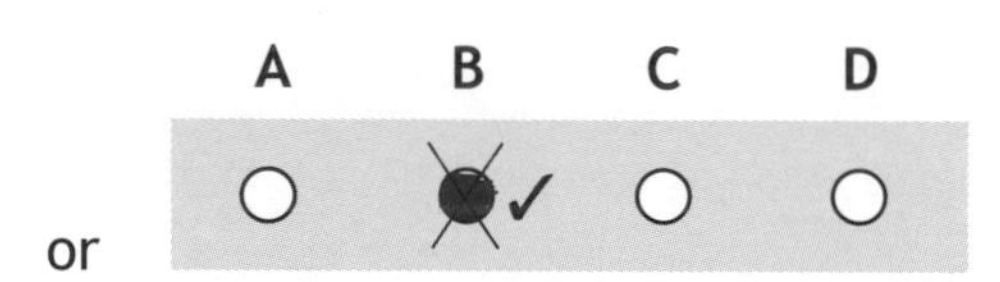

SECTION 1 — Answer Grid

	A	B	C	D	E
1	○	○	○	○	○
2	○	○	○	○	○
3	○	○	○	○	○
4	○	○	○	○	○
5	○	○	○	○	○
6	○	○	○	○	○
7	○	○	○	○	○
8	○	○	○	○	○
9	○	○	○	○	○
10	○	○	○	○	○
11	○	○	○	○	○
12	○	○	○	○	○
13	○	○	○	○	○
14	○	○	○	○	○
15	○	○	○	○	○
16	○	○	○	○	○
17	○	○	○	○	○
18	○	○	○	○	○
19	○	○	○	○	○
20	○	○	○	○	○

[BLANK PAGE]

SECTION 2 — 90 marks

Attempt ALL questions

MARKS | DO NOT WRITE IN THIS MARGIN

1. A student reproduces Galileo's famous experiment by dropping a solid copper ball of mass 0·50 kg from a balcony on the Leaning Tower of Pisa.

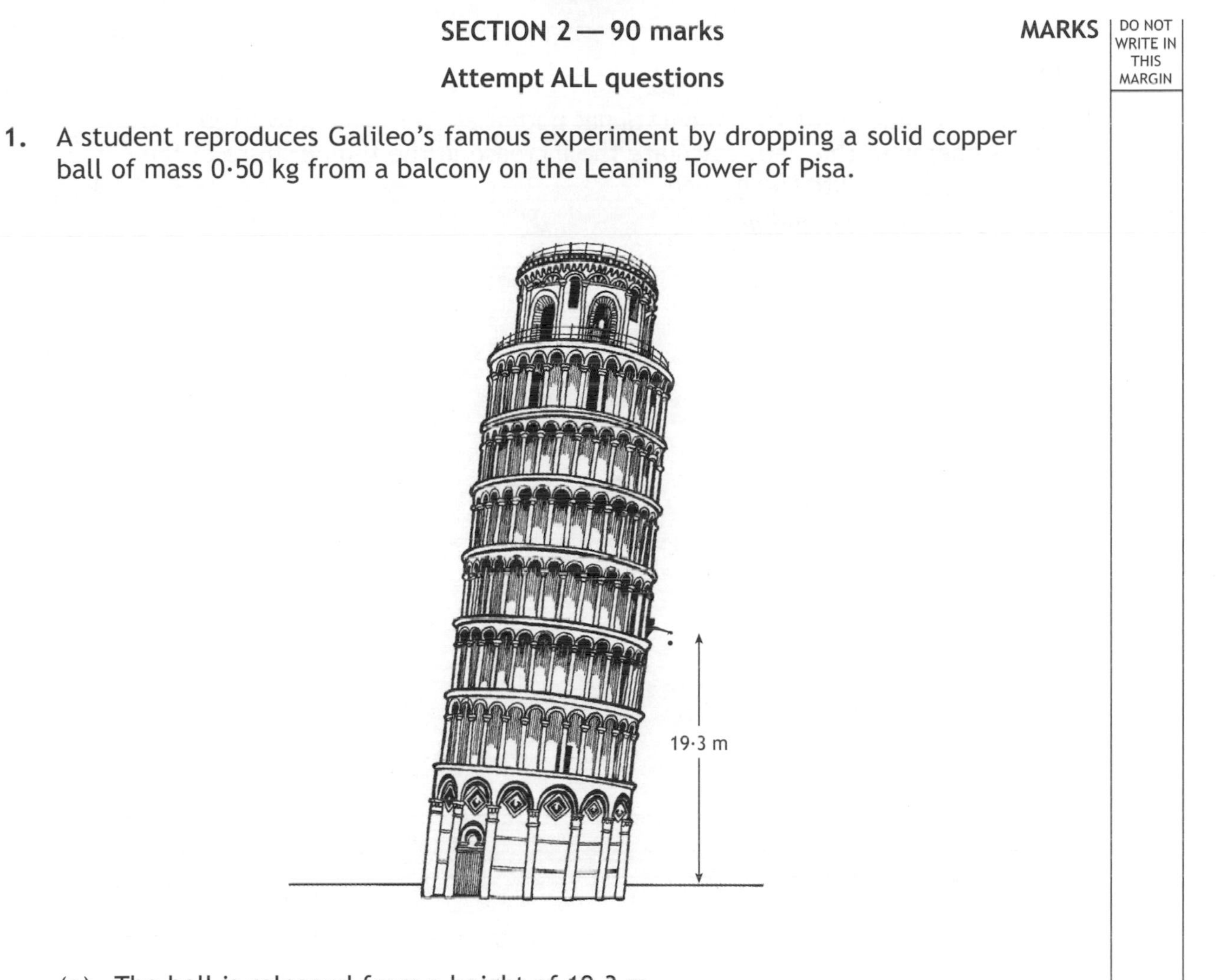

(a) The ball is released from a height of 19·3 m.

Calculate the gravitational potential energy lost by the ball. **3**

Space for working and answer.

MARKS | DO NOT WRITE IN THIS MARGIN

1. (continued)

(b) Assuming that all of this gravitational potential energy is converted into heat energy **in the ball**, calculate the increase in the temperature of the ball on impact with the ground. **3**

Space for working and answer.

(c) Is the actual temperature change of the ball greater than, the same as or less than the value calculated in part (a)(ii)?

You **must** explain your answer. **2**

Total marks 8

MARKS DO NOT WRITE IN THIS MARGIN

2. Some resistors are labelled with a power rating as well as their resistance value. This is the maximum power at which they can operate without overheating.

(a) A resistor is labelled 50 Ω, 2W.

Calculate the maximum operating current for this resistor. **3**

Space for working and answer.

(b) Two resistors, each rated at 2W, are connected in parallel to a 9 V d.c. supply.

They have resistances of 60 Ω and 30 Ω.

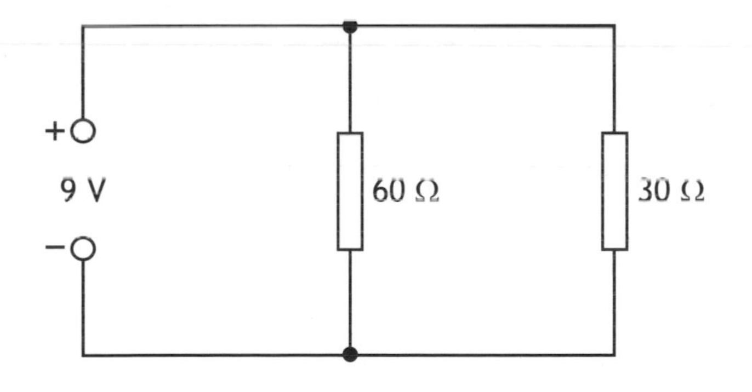

MARKS | DO NOT WRITE IN THIS MARGIN

2. (continued)

(i) Calculate the total resistance of the circuit. **3**

Space for working and answer.

(ii) Calculate the power produced in each resistor. **4**

Space for working and answer.

(iii) State which, if any, of the resistors will overheat. **1**

(c) The 9 V **d.c.** supply is replaced by a 9 V **a.c.** supply.

What effect, if any, would this have on your answers to part (b) (ii)? **1**

Total marks 12

MARKS DO NOT WRITE IN THIS MARGIN

3. A mass of copper heated with a Bunsen is immersed in a beaker of cold water.

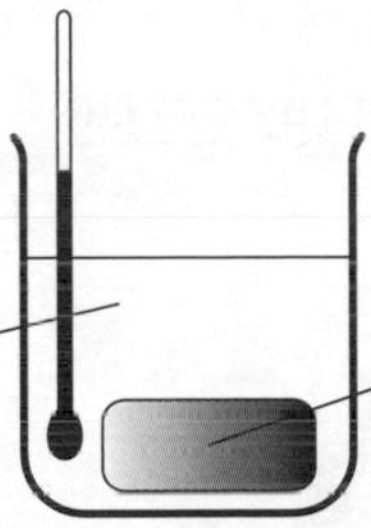

water
copper

Use your knowledge of physics to comment on what the final temperature of the copper and water would depend on.

Make reference to any relevant equation(s) in your answer. **3**

MARKS | DO NOT WRITE IN THIS MARGIN

4. A garden spray consists of a tank, a pump and a spray nozzle.

The tank is partially filled with water.

The pump is then used to increase the pressure of the air above the water.

The volume of the compressed air in the tank is $1{\cdot}60 \times 10^{-3}$ m^3.

The surface area of the water is $3{\cdot}00 \times 10^{-2}$ m^2.

The pressure of the air in the tank is $4{\cdot}60 \times 10^5$ Pa.

(a) Calculate the force on the surface of the water. **3**

Space for working and answer.

MARKS DO NOT WRITE IN THIS MARGIN

4. (continued)

(b) The spray nozzle is operated and water is pushed out until the pressure of the air in the tank is $1 \cdot 00 \times 10^5$ Pa.

Calculate the new volume of compressed air in the tank. **3**

Space for working and answer.

(c) Calculate the volume of water expelled. **1**

Space for working and answer.

Total marks 7

MARKS | DO NOT WRITE IN THIS MARGIN

5. A ripple tank is set up to investigate the properties of water waves.

A wave generator is used to produce the waves in the tank.

(a) When the wave generator is vibrating at 5 Hz, it is found that there are 8 complete waves between the wave generator and the opposite side of the tank, as shown in figure 1.

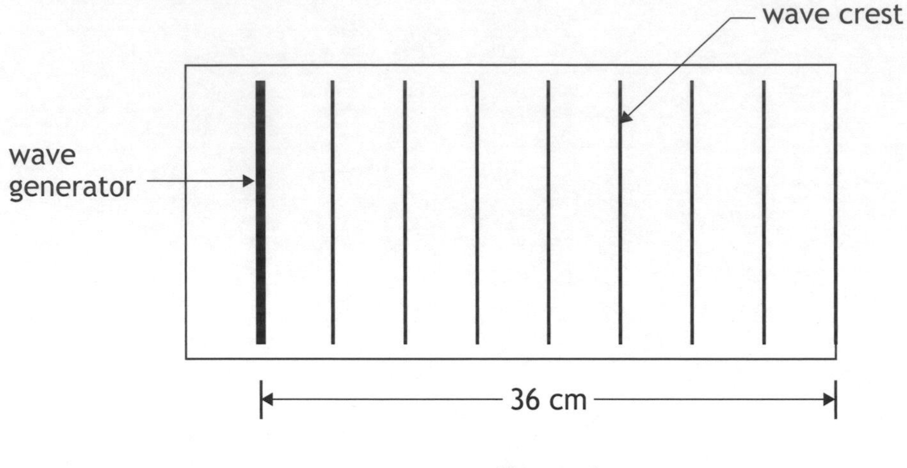

figure 1

Calculate the speed of the water waves. **4**

Space for working and answer.

MARKS | DO NOT WRITE IN THIS MARGIN

5. (continued)

(b) A barrier with a wide gap in it is placed across the middle of the tank as shown in figure 2.

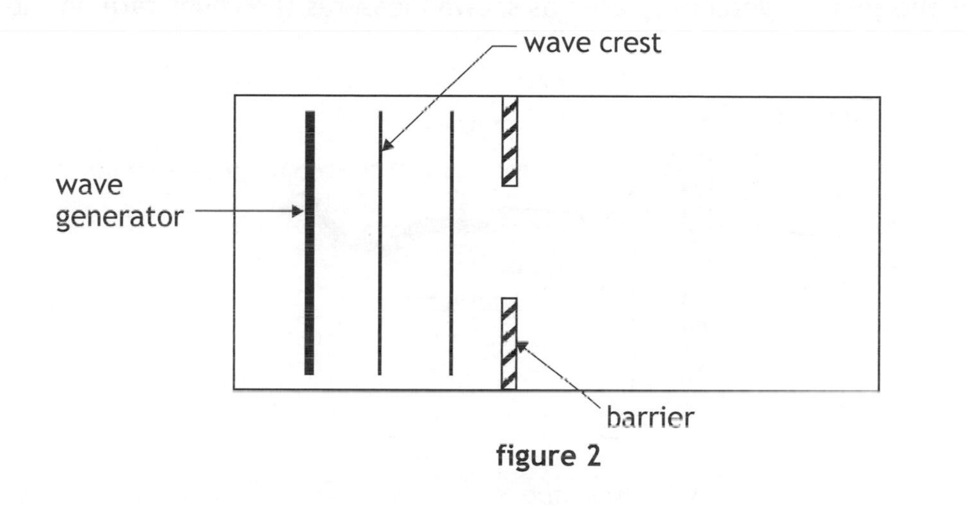

figure 2

Complete figure 2 showing the wave pattern on the right-hand side of the barrier. **2**

(c) Optical fibres are used to carry internet data using infra-red radiation.

The diagram shows a view of an infra-red ray entering the end of a fibre.

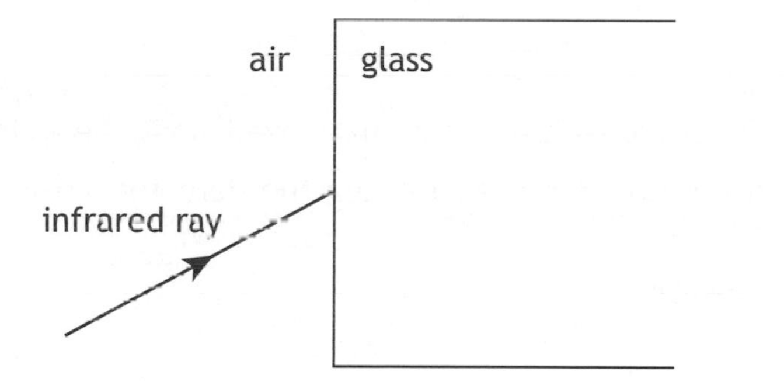

Complete the diagram to show the path of the infra-red ray as it enters the glass from air.

Indicate on your diagram the normal, the angle of incidence and the angle of refraction. **2**

Total marks 8

MARKS DO NOT WRITE IN THIS MARGIN

6. When welders join thick steel plates it is important that the joint is completely filled with metal. This ensures there are no air pockets in the metal weld, as this would weaken the joint.

One method of checking for air pockets is to use a radioactive source on one side of the joint. A detector placed as shown measures the count rate on the other side.

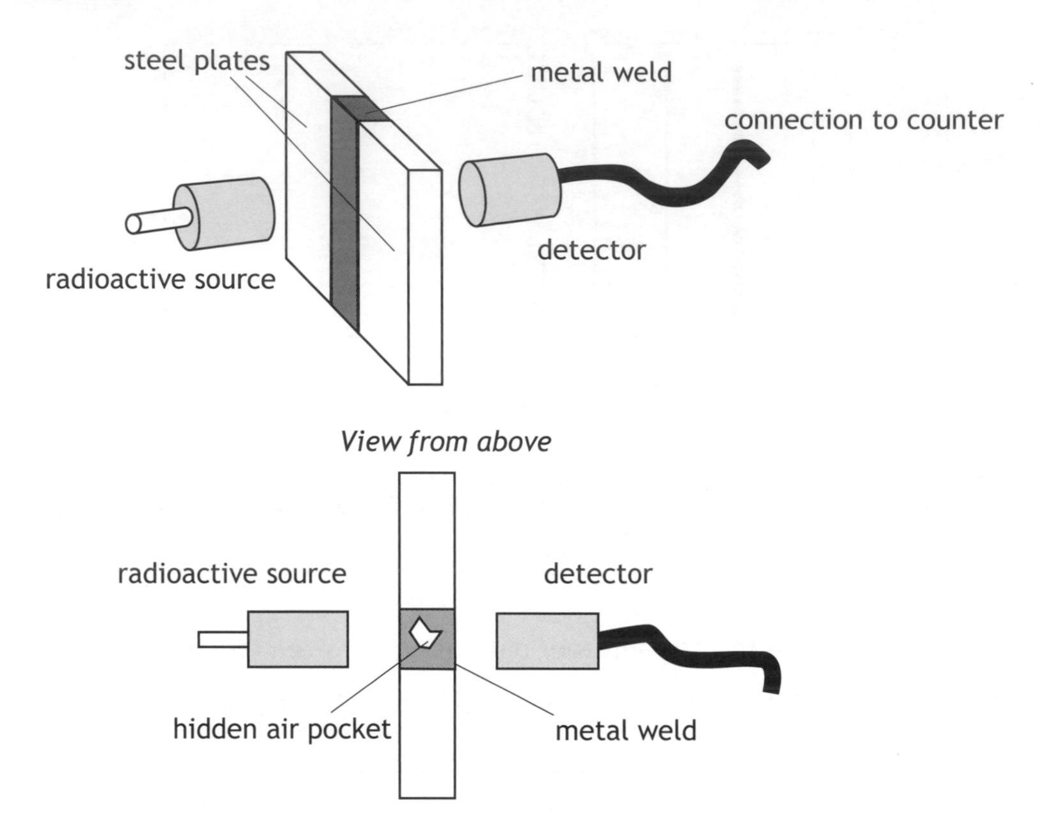

(a) The radioactive source and detector are moved along the weld.

How would the count rate change when the detector moves over an air pocket?

Explain your answer. **2**

(b) Which of the radiations alpha, beta or gamma must be used?

Explain your answer. **2**

Total marks 4

MARKS DO NOT WRITE IN THIS MARGIN

7. (a) A medical physicist checks the count rate of a radioactive source. A graph of count rate against time for the source is shown. The count rate has been corrected for background radiation.

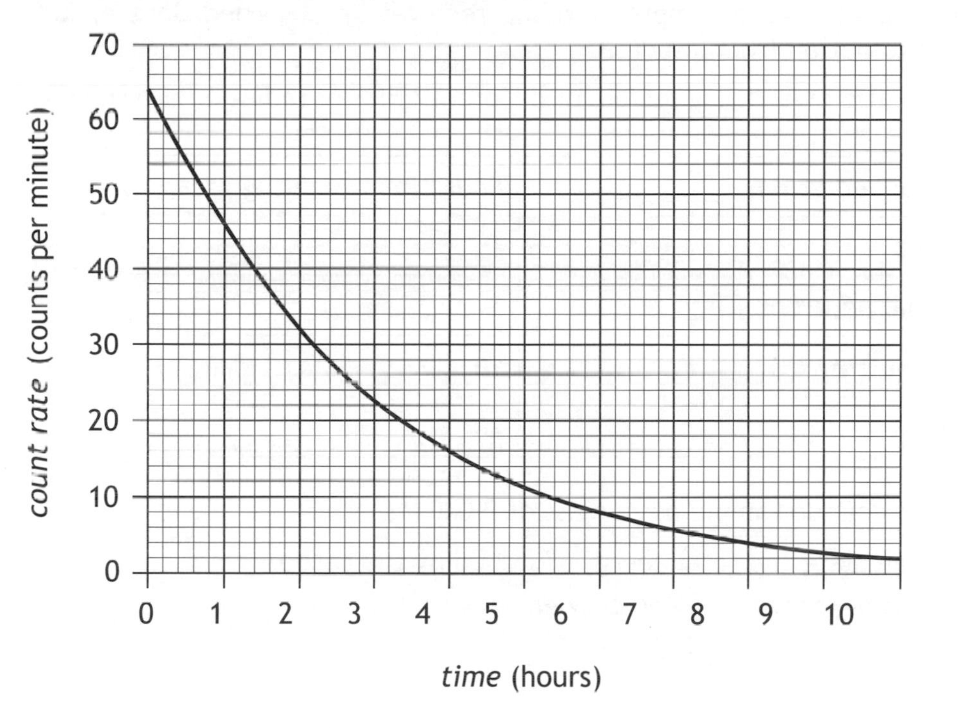

(i) Use the graph to determine the half-life of the source. **1**

(ii) State **two** factors which can affect the background radiation level. **2**

MARKS | DO NOT WRITE IN THIS MARGIN

7. (continued)

(b) Another medical physicist is investigating the effects of radiation on tissue samples. One sample of tissue receives an absorbed dose of 500 μGy of radiation from a source.

The radiation weighting factors of different types of radiation are shown.

Type of radiation	*Radiation weighting factor* (w_R)
gamma	1
thermal neutrons	3
fast neutrons	10
alpha	20

(i) The tissue sample has a mass of 0·040 kg.

Calculate the total energy absorbed by the tissue. **3**

Space for working and answer.

(ii) The equivalent dose received for this tissue sample is 10·00 mSv.

Which type of radiation is the medical physicist using?

Justify your answer by calculation. **4**

Total marks 10

MARKS DO NOT WRITE IN THIS MARGIN

8. A simple pendulum consists of a mass suspended on a string which is allowed to swing from a point. The pendulum oscillates back and forth when released.

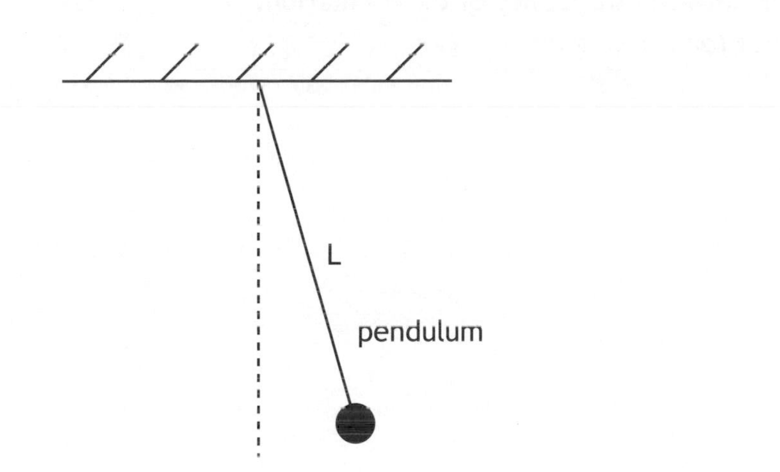

For small amplitudes, the period T of this pendulum can be approximated by the relationship:

$$T = 2\pi\sqrt{\frac{L}{g}}$$

Where: L is the length of the string (m).

g is the acceleration due to gravity (m s^{-2}).

(a) Calculate the period of a pendulum with a string of length 0·8 m. **2**

Space for working and answer.

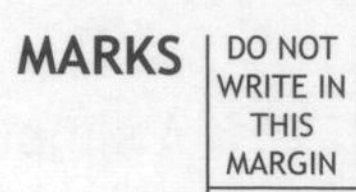

8. (continued)

(b) Calculate the frequency of the oscillation. **3**

Space for working and answer.

(c) The mass of the pendulum in the above example is now doubled.

State the effect this has on the period of the pendulum. **1**

Total marks 6

MARKS DO NOT WRITE IN THIS MARGIN

9. The European Space Agency has been authorised to fly manned missions to the International Space Station (ISS).

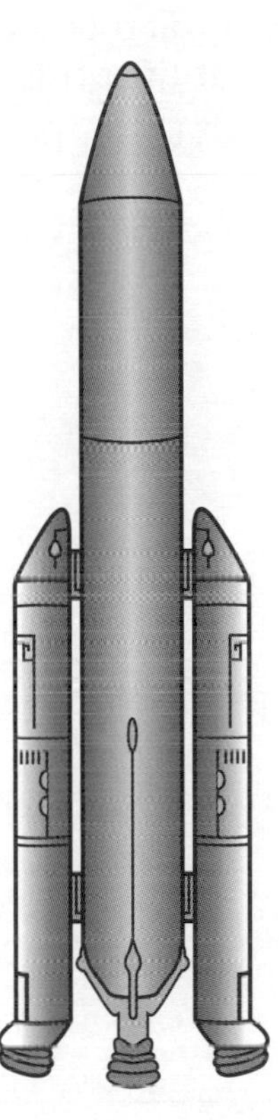

For one particular mission, a spacecraft with booster rockets attached will be launched.

(a) On the diagram above draw and label the two forces acting on the spacecraft at lift-off. **2**

MARKS | DO NOT WRITE IN THIS MARGIN

9. (continued)

(b) The combined mass of the spacecraft and booster rockets is $3{\cdot}08 \times 10^5$ kg and the initial thrust on the rocket at lift-off is 3352 kN.

The frictional forces acting on the rocket at lift-off are negligible.

(i) Calculate the weight of the spacecraft and booster rockets at lift-off. **3**

Space for working and answer.

(ii) Calculate the acceleration of the spacecraft and booster rockets at lift-off. **4**

Space for working and answer.

MARKS DO NOT WRITE IN THIS MARGIN

9. (continued)

(c) The ISS orbits at a height of approximately 360 km above the Earth.

Explain why the ISS stays in orbit around the Earth. **2**

(d) An astronaut on board the ISS takes part in a video link-up with a group of students. The students see the astronaut floating.

(i) Explain why the astronaut appears to float. **1**

(ii) The astronaut then pushes against a wall and moves off.

Explain in terms of Newton's Third Law why the astronaut moves. **2**

Total marks 14

MARKS DO NOT WRITE IN THIS MARGIN

10. A car of mass 700 kg travels along a motorway at a constant speed. The driver sees a traffic hold-up ahead and performs an emergency stop. A graph of the car's motion is shown, from the moment the driver sees the hold-up.

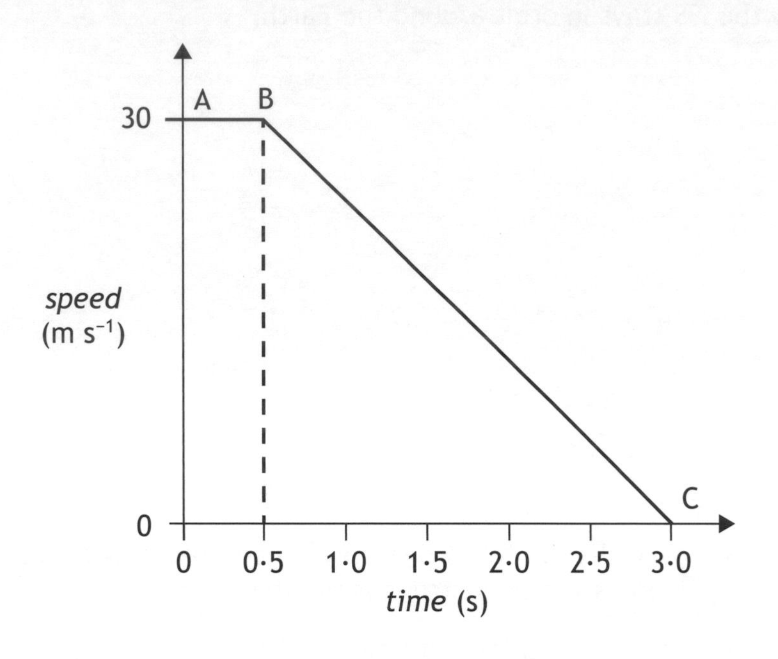

(a) Describe **and** explain the motion of the car between A and B. **2**

MARKS DO NOT WRITE IN THIS MARGIN

10. (continued)

(b) Calculate the kinetic energy of the car at A. 3

Space for working and answer.

(c) Show by calculation that the magnitude of the unbalanced force required to bring the car to a halt between B and C is 8400 N. 4

Space for working and answer.

Total marks 9

MARKS | DO NOT WRITE IN THIS MARGIN

11. Some information about two racing cars is shown in the table.

Car	Maximum speed ($m\ s^{-1}$)	maximum acceleration ($m\ s^{-2}$)
A	40	4
B	20	9

The cars race on the following track:

Use your knowledge of physics to comment on which car would be most suitable for racing on this track. **3**

MARKS DO NOT WRITE IN THIS MARGIN

12. Read the passage below and answer the questions that follow.

Sunspots

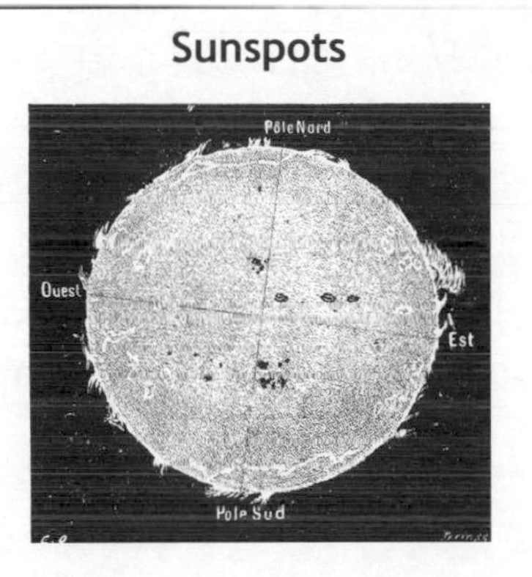

Sunspots are the dark spots which appear on the surface of the sun. They are caused by intense magnetic fields appearing beneath the sun's surface. These magnetic fields disrupt the natural convection within the sun, causing a reduction in the surface temperature at that point which appears as a dark spot.

Sunspots release huge amounts of energy known as "solar flares". These flares lead to the ejection of billions of tonnes of high-energy solar particles. Electromagnetic (EM) radiation is also emitted. Such releases are known as coronal mass ejections (CME). The particles and energy travel into space.

High energy particles emitted by solar flares rarely reach Earth, and when they do, its magnetic field deflects most of them. Background radiation levels are not significantly changed by the particles which penetrate the atmosphere.

However, satellites in the path of these particles can suffer damage to sensitive equipment on board. Early warning of the arrival of solar flares allows operators to place satellites into "protective mode".

The arrival of high electromagnetic energy, in particular ultraviolet radiation and X-rays, can also cause significant damage.

When this EM radiation reaches the Earth, it can cause changes to the upper atmosphere, which can affect the orbits of low-altitude satellites. The EM radiation can also cause localized disturbances of the Earth's magnetic field. On one occasion, such interference was claimed to be responsible for causing power failures in Canada which affected millions of households for several hours.

The NASA space agency monitors sunspots, using their two satellite-based Solar Terrestrial Relations Observatories (STEREO).

One particular observation from these satellites recorded a solar flare ejected from the sun which took 2·3 days to reach Mars, a distance of $2{\cdot}28\times10^{11}$ metres from the sun.

Sunspot activity increases to a maximum in 11-year cycles.

There is on-going research to identify a connection between solar activity and our terrestrial climate.

Currently, scientists cannot predict when a sunspot or solar flare will appear, but detection and early warning techniques have improved.

MARKS | DO NOT WRITE IN THIS MARGIN

12. (continued)

(a) What causes sunspots to appear? 1

(b) Which parts of the electromagnetic spectrum can cause damage to satellites? 2

(c) Calculate the average speed in $m\,s^{-1}$ of the solar flare which reached Mars. 3

Space for working and answer.

Total marks 6

[END OF MODEL PAPER]

MARKS DO NOT WRITE IN THIS MARGIN

ADDITIONAL SPACE FOR ROUGH WORKING AND ANSWERS

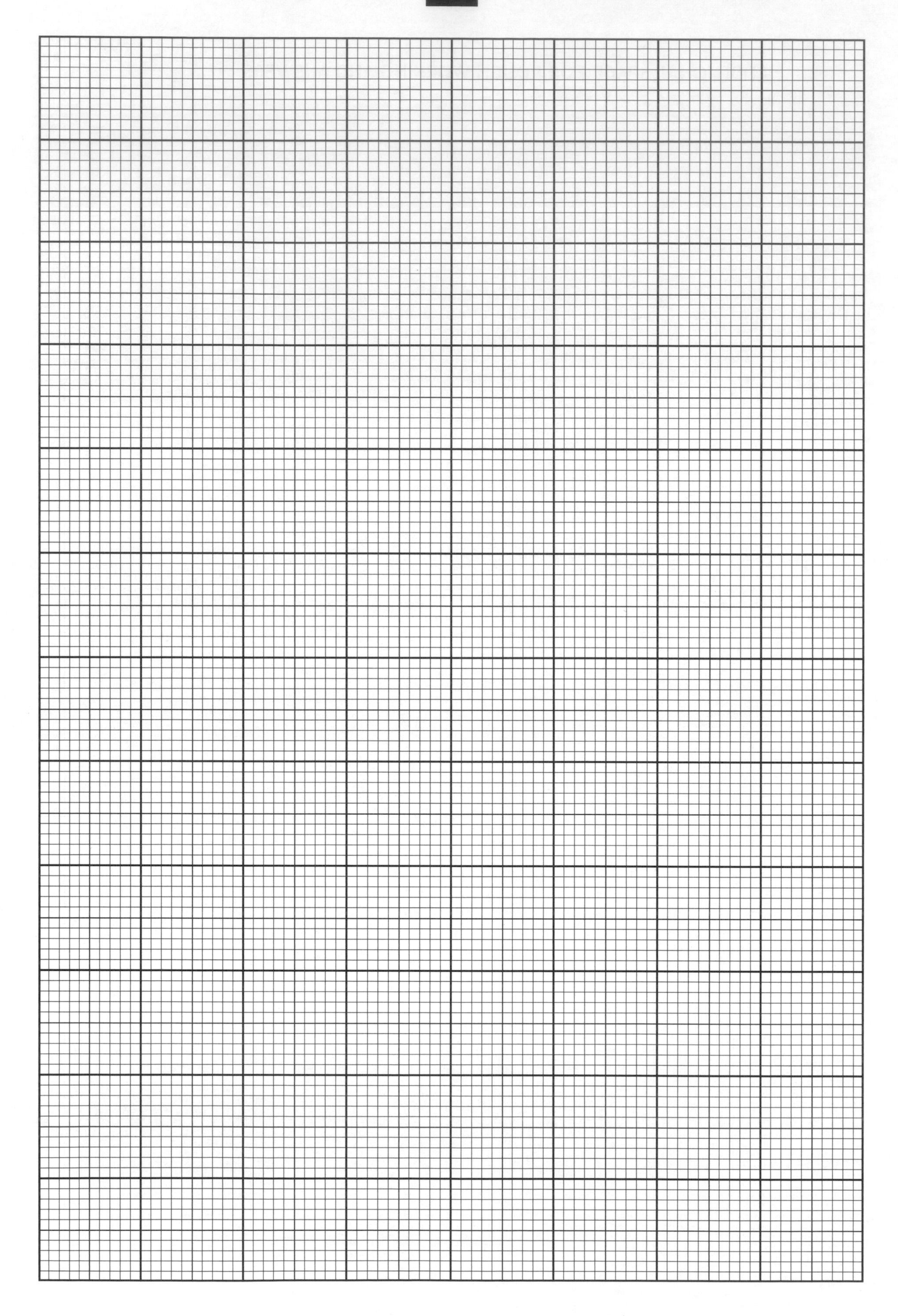

NATIONAL 5
2014

SQA
©

HODDER
GIBSON
LEARN MORE

National
Qualifications
2014

X757/75/02

Physics
Section 1—Questions

THURSDAY, 22 MAY

9:00 AM – 11:00 AM

Instructions for the completion of Section 1 are given on Page two of your question and answer booklet X757/75/01.

Record your answers on the answer grid on Page three of your question and answer booklet.

Reference may be made to the Data Sheet on Page two of this booklet and to the Relationship Sheet X757/75/11.

Before leaving the examination room you must give your question and answer booklet to the Invigilator; if you do not, you may lose all the marks for this paper.

DATA SHEET

Speed of light in materials

Material	*Speed in* m s^{-1}
Air	$3{\cdot}0 \times 10^8$
Carbon dioxide	$3{\cdot}0 \times 10^8$
Diamond	$1{\cdot}2 \times 10^8$
Glass	$2{\cdot}0 \times 10^8$
Glycerol	$2{\cdot}1 \times 10^8$
Water	$2{\cdot}3 \times 10^8$

Gravitational field strengths

	Gravitational field strength on the surface in N kg^{-1}
Earth	9·8
Jupiter	23
Mars	3·7
Mercury	3·7
Moon	1·6
Neptune	11
Saturn	9·0
Sun	270
Uranus	8·7
Venus	8·9

Specific latent heat of fusion of materials

Material	*Specific latent heat of fusion in* J kg^{-1}
Alcohol	$0{\cdot}99 \times 10^5$
Aluminium	$3{\cdot}95 \times 10^5$
Carbon dioxide	$1{\cdot}80 \times 10^5$
Copper	$2{\cdot}05 \times 10^5$
Iron	$2{\cdot}67 \times 10^5$
Lead	$0{\cdot}25 \times 10^5$
Water	$3{\cdot}34 \times 10^5$

Specific latent heat of vaporisation of materials

Material	*Specific latent heat of vaporisation in* J kg^{-1}
Alcohol	$11{\cdot}2 \times 10^5$
Carbon dioxide	$3{\cdot}77 \times 10^5$
Glycerol	$8{\cdot}30 \times 10^5$
Turpentine	$2{\cdot}90 \times 10^5$
Water	$22{\cdot}6 \times 10^5$

Speed of sound in materials

Material	*Speed in* m s^{-1}
Aluminium	5200
Air	340
Bone	4100
Carbon dioxide	270
Glycerol	1900
Muscle	1600
Steel	5200
Tissue	1500
Water	1500

Specific heat capacity of materials

Material	*Specific heat capacity in* J kg^{-1} °C^{-1}
Alcohol	2350
Aluminium	902
Copper	386
Glass	500
Ice	2100
Iron	480
Lead	128
Oil	2130
Water	4180

Melting and boiling points of materials

Material	*Melting point in* °C	*Boiling point in* °C
Alcohol	−98	65
Aluminium	660	2470
Copper	1077	2567
Glycerol	18	290
Lead	328	1737
Iron	1537	2737

Radiation weighting factors

Type of radiation	*Radiation weighting factor*
alpha	20
beta	1
fast neutrons	10
gamma	1
slow neutrons	3
X-rays	1

SECTION 1

1. The voltage of an electrical supply is a measure of the

A resistance of the circuit
B speed of the charges in the circuit
C power developed in the circuit
D energy given to the charges in the circuit
E current in the circuit.

2. Four circuit symbols, W, X, Y and Z, are shown.

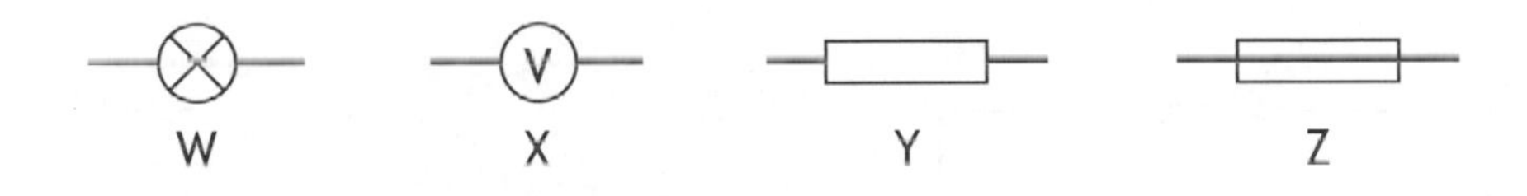

Which row identifies the components represented by these symbols?

	W	X	Y	Z
A	battery	ammeter	resistor	variable resistor
B	battery	ammeter	fuse	resistor
C	lamp	ammeter	variable resistor	resistor
D	lamp	voltmeter	resistor	fuse
E	lamp	voltmeter	variable resistor	fuse

[Turn over

3. A student suspects that ammeter A_1 may be inaccurate. Ammeter A_2 is known to be accurate.

Which of the following circuits should be used to compare the reading on A_1 with A_2?

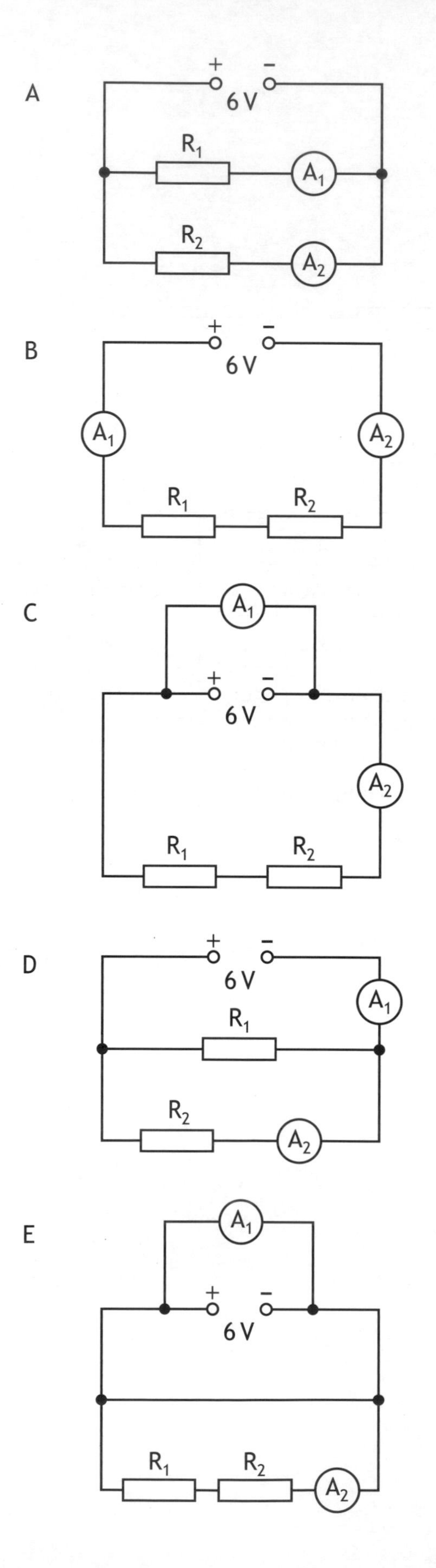

4. A ball of mass 0·50 kg is released from a height of 1·00 m and falls towards the floor.

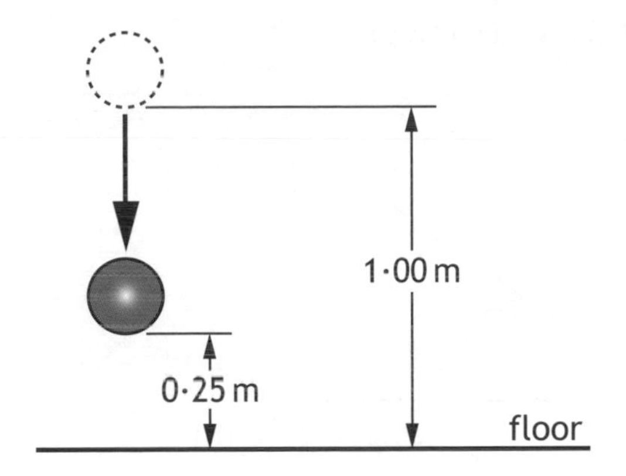

Which row in the table shows the gravitational potential energy and the kinetic energy of the ball when it is at a height of 0·25 m from the floor?

	Gravitational potential energy (J)	*Kinetic energy* (J)
A	0·12	0·12
B	1·2	1·2
C	1·2	3·7
D	3·7	1·2
F	4·9	1·2

5. The pressure of a fixed mass of gas is $6{\cdot}0 \times 10^5$ Pa.

The temperature of the gas is 27 °C and the volume of the gas is $2{\cdot}5\ m^3$.

The temperature of the gas increases to 54 °C and the volume of the gas increases to $5{\cdot}0\ m^3$.

What is the new pressure of the gas?

A $2{\cdot}8 \times 10^5$ Pa

B $3{\cdot}3 \times 10^5$ Pa

C $6{\cdot}0 \times 10^5$ Pa

D $1{\cdot}1 \times 10^6$ Pa

E $1{\cdot}3 \times 10^6$ Pa

[Turn over

6. A student is investigating the relationship between the volume and the kelvin temperature of a fixed mass of gas at constant pressure.

Which graph shows this relationship?

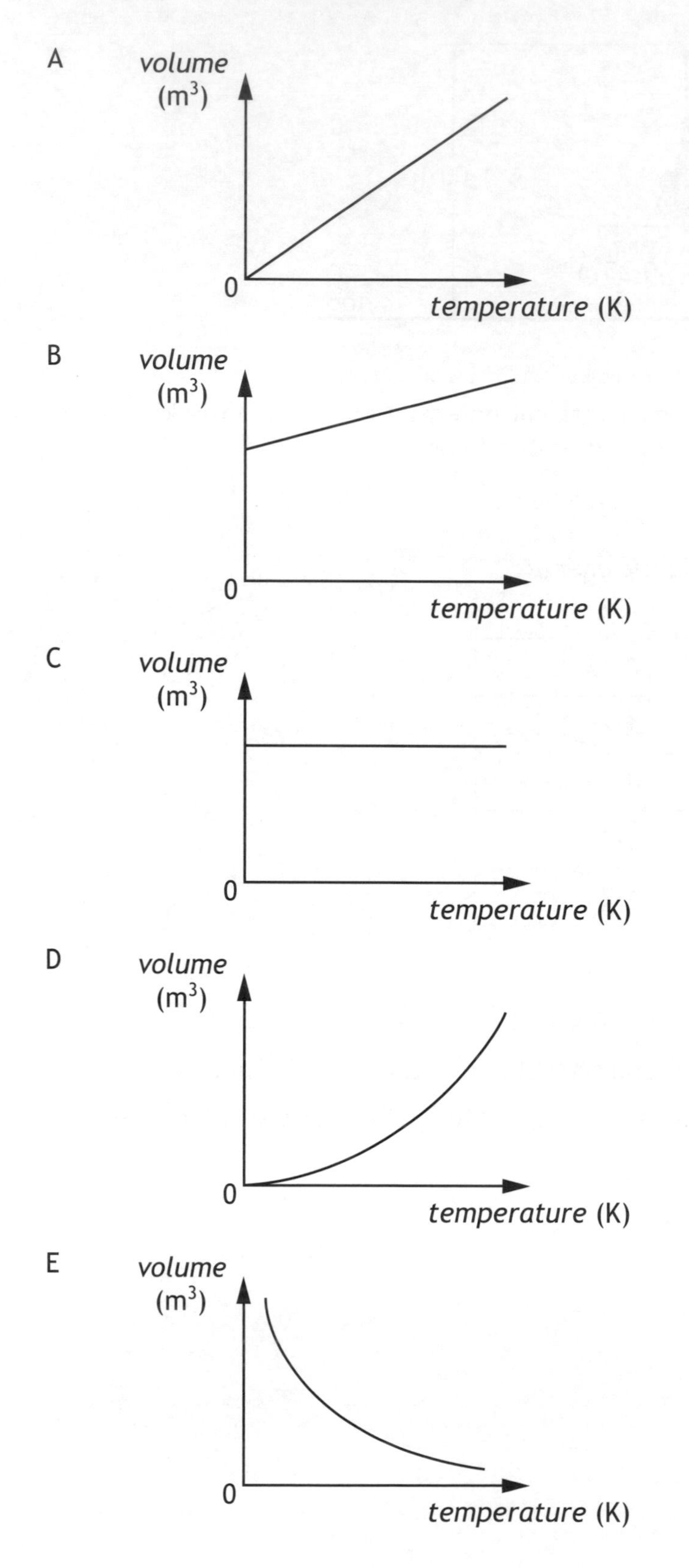

7. A liquid is heated from 17 °C to 50 °C. The temperature rise in kelvin is

A 33 K
B 67 K
C 306 K
D 340 K
E 579 K.

8. The period of vibration of a guitar string is 8 ms.

The frequency of the sound produced by the guitar string is

A 0·125 Hz
B 12·5 Hz
C 125 Hz
D 800 Hz
E 8000 Hz.

9. A student makes the following statements about microwaves and radio waves.

I In air, microwaves travel faster than radio waves.
II In air, microwaves have a longer wavelength than radio waves.
III Microwaves and radio waves are both members of the electromagnetic spectrum.

Which of these statements is/are correct?

A I only
B III only
C I and II only
D I and III only
E II and III only

10. Which row describes alpha (α), beta (β) and gamma (γ) radiations?

	α	β	γ
A	helium nucleus	electromagnetic radiation	electron from the nucleus
B	helium nucleus	electron from the nucleus	electromagnetic radiation
C	electron from the nucleus	helium nucleus	electromagnetic radiation
D	electromagnetic radiation	helium nucleus	electron from the nucleus
E	electromagnetic radiation	electron from the nucleus	helium nucleus

[Turn over

11. A sample of tissue is irradiated using a radioactive source.

A student makes the following statements about the sample.

I The equivalent dose received by the sample is reduced by shielding the sample with a lead screen.

II The equivalent dose received by the sample is increased as the distance from the source to the sample is increased.

III The equivalent dose received by the sample is increased by increasing the time of exposure of the sample to the radiation.

Which of these statements is/are correct?

A I only

B II only

C I and II only

D II and III only

E I and III only

12. The half-life of a radioactive source is 64 years.

In 2 hours, $1{\cdot}44 \times 10^8$ radioactive nuclei in the source decay.

What is the activity of the source in Bq?

A 2×10^4

B 4×10^4

C $1{\cdot}2 \times 10^6$

D $2{\cdot}25 \times 10^6$

E $7{\cdot}2 \times 10^7$

13. A student makes the following statements about the fission process in a nuclear power station.

I Electrons are used to bombard a uranium nucleus.

II Heat is produced.

III The neutrons released can cause other nuclei to undergo fission.

Which of these statements is/are correct?

A I only

B II only

C III only

D I and II only

E II and III only

14. Which of the following contains two vectors and one scalar quantity?

A Acceleration, mass, displacement

B Displacement, force, velocity

C Time, distance, force

D Displacement, velocity, acceleration

E Speed, velocity, distance

15. A vehicle follows a course from R to T as shown.

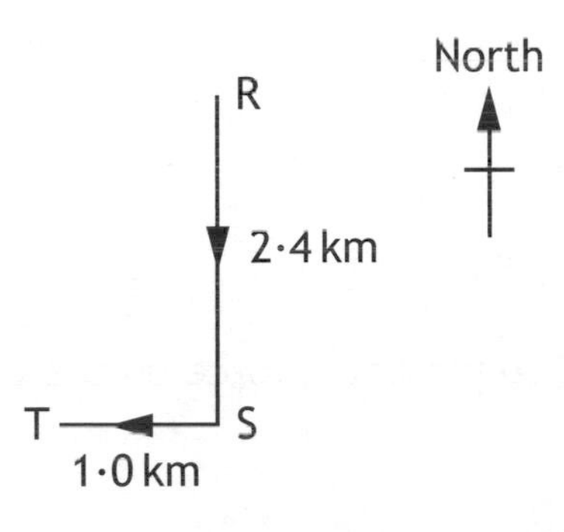

The total journey takes 1 hour.

Which row in the table gives the average speed and the average velocity of the vehicle for the whole journey?

	Average speed	*Average velocity*
A	$2{\cdot}6\ \text{km h}^{-1}$ (023)	$3{\cdot}4\ \text{km h}^{-1}$
B	$2{\cdot}6\ \text{km h}^{-1}$	$3{\cdot}4\ \text{km h}^{-1}$ (203)
C	$3{\cdot}4\ \text{km h}^{-1}$ (203)	$2{\cdot}6\ \text{km h}^{-1}$
D	$3{\cdot}4\ \text{km h}^{-1}$	$2{\cdot}6\ \text{km h}^{-1}$ (023)
E	$3{\cdot}4\ \text{km h}^{-1}$	$2{\cdot}6\ \text{km h}^{-1}$ (203)

16. A force of 10 N acts on an object for 2 s.

During this time the object moves a distance of 3 m.

The work done on the object is

A 6·7 J

B 15 J

C 20 J

D 30 J

E 60 J.

17. Catapults are used by anglers to project fish bait into water.

A technician designs a catapult for this use.

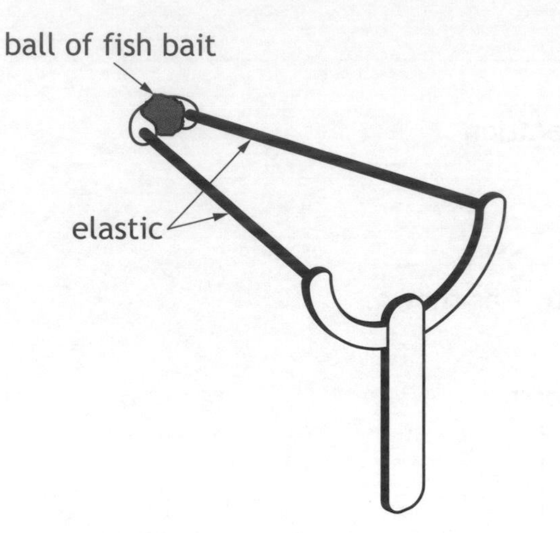

Pieces of elastic of different thickness are used to provide a force on the ball.

Each piece of elastic is the same length.

The amount of stretch given to each elastic is the same each time.

The force exerted on the ball increases as the thickness of the elastic increases.

Which row in the table shows the combination of the thickness of elastic and mass of ball that produces the greatest acceleration?

	Thickness of elastic (mm)	*Mass of ball* (kg)
A	5	0·01
B	10	0·01
C	10	0·02
D	15	0·01
E	15	0·02

18. A spacecraft completes the last stage of its journey back to Earth by parachute, falling with constant speed into the sea.

The spacecraft falls with constant speed because

A the gravitational field strength of the Earth is constant near the Earth's surface

B it has come from space where the gravitational field strength is almost zero

C the air resistance is greater than the weight of the spacecraft

D the weight of the spacecraft is greater than the air resistance

E the air resistance is equal to the weight of the spacecraft.

19. A ball is released from point **Q** on a curved rail, leaves the rail horizontally at R and lands 1 s later.

The ball is now released from point **P**.

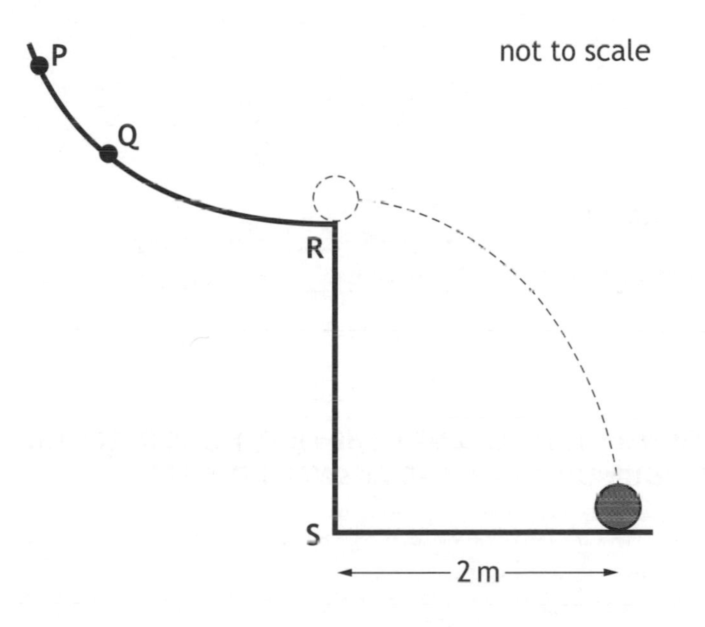

Which row describes the motion of the ball after leaving the rail?

	Time to land after leaving rail	*Distance from S to landing point*
A	1 s	less than 2 m
B	less than 1 s	more than 2 m
C	1 s	more than 2 m
D	less than 1 s	2 m
E	more than 1 s	more than 2 m

20. A solid substance is placed in an insulated flask and heated continuously with an immersion heater.

The graph shows how the temperature of the substance in the flask changes in time.

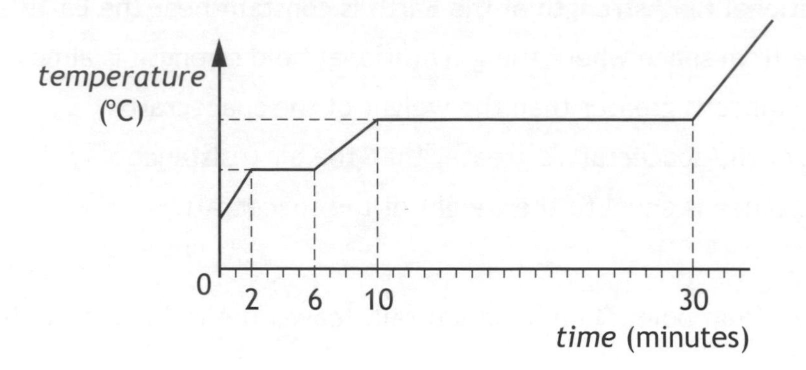

After 5 minutes the substance is a

A solid

B liquid

C gas

D mixture of solid and liquid

E mixture of liquid and gas.

[END OF SECTION 1. NOW ATTEMPT THE QUESTIONS IN SECTION 2 OF YOUR QUESTION AND ANSWER BOOKLET]

National
Qualifications
2014

X757/75/11

Physics
Relationships Sheet

THURSDAY, 22 MAY
9:00 AM – 11:00 AM

$E_p = mgh$

$E_k = \frac{1}{2}mv^2$

$Q = It$

$V = IR$

$R_T = R_1 + R_2 + \ldots$

$\frac{1}{R_T} = \frac{1}{R_1} + \frac{1}{R_2} + \ldots$

$V_2 = \left(\frac{R_2}{R_1 + R_2}\right)V_s$

$\frac{V_1}{V_2} = \frac{R_1}{R_2}$

$P = \frac{E}{t}$

$P = IV$

$P = I^2R$

$P = \frac{V^2}{R}$

$E_h = cm\Delta T$

$p = \frac{F}{A}$

$\frac{pV}{T} = \text{constant}$

$p_1V_1 = p_2V_2$

$\frac{p_1}{T_1} = \frac{p_2}{T_2}$

$\frac{V_1}{T_1} = \frac{V_2}{T_2}$

$d = vt$

$v = f\lambda$

$T = \frac{1}{f}$

$A = \frac{N}{t}$

$D = \frac{E}{m}$

$H = Dw_R$

$\dot{H} = \frac{H}{t}$

$s = vt$

$d = \bar{v}t$

$s = \bar{v}t$

$a = \frac{v-u}{t}$

$W = mg$

$F = ma$

$E_w = Fd$

$E_h = ml$

Additional Relationships

Circle

circumference $= 2\pi r$

area $= \pi r^2$

Sphere

area $= 4\pi r^2$

volume $= \frac{4}{3}\pi r^3$

Trigonometry

$$\sin\Theta = \frac{\text{opposite}}{\text{hypotenuse}}$$

$$\cos\Theta = \frac{\text{adjacent}}{\text{hypotenuse}}$$

$$\tan\Theta = \frac{\text{opposite}}{\text{adjacent}}$$

$$\sin^2\Theta + \cos^2\Theta = 1$$

Electron Arrangements of Elements

Key

Atomic number Symbol Electron arrangement Name

Group 1 (1)	Group 2 (2)	(3)	(4)	(5)	(6)	(7)	(8)	(9)	(10)	(11)	(12)	Group 3 (13)	Group 4 (14)	Group 5 (15)	Group 6 (16)	Group 7 (17)	Group 0 (18)
1 **H** 1 Hydrogen																	2 **He** 2 Helium
3 **Li** 2,1 Lithium	4 **Be** 2,2 Beryllium											5 **B** 2,3 Boron	6 **C** 2,4 Carbon	7 **N** 2,5 Nitrogen	8 **O** 2,6 Oxygen	9 **F** 2,7 Fluorine	10 **Ne** 2,8 Neon
11 **Na** 2,8,1 Sodium	12 **Mg** 2,8,2 Magnesium											13 **Al** 2,8,3 Aluminium	14 **Si** 2,8,4 Silicon	15 **P** 2,8,5 Phosphorus	16 **S** 2,8,6 Sulfur	17 **Cl** 2,8,7 Chlorine	18 **Ar** 2,8,8 Argon
19 **K** 2,8,8,1 Potassium	20 **Ca** 2,8,8,2 Calcium	21 **Sc** 2,8,9,2 Scandium	22 **Ti** 2,8,10,2 Titanium	23 **V** 2,8,11,2 Vanadium	24 **Cr** 2,8,13,1 Chromium	25 **Mn** 2,8,13,2 Manganese	26 **Fe** 2,8,14,2 Iron	27 **Co** 2,8,15,2 Cobalt	28 **Ni** 2,8,16,2 Nickel	29 **Cu** 2,8,18,1 Copper	30 **Zn** 2,8,18,2 Zinc	31 **Ga** 2,8,18,3 Gallium	32 **Ge** 2,8,18,4 Germanium	33 **As** 2,8,18,5 Arsenic	34 **Se** 2,8,18,6 Selenium	35 **Br** 2,8,18,7 Bromine	36 **Kr** 2,8,18,8 Krypton
37 **Rb** 2,8,18,8,1 Rubidium	38 **Sr** 2,8,18,8,2 Strontium	39 **Y** 2,8,18,9,2 Yttrium	40 **Zr** 2,8,18, 10,2 Zirconium	41 **Nb** 2,8,18, 12,1 Niobium	42 **Mo** 2,8,18,13, 1 Molybdenum	43 **Tc** 2,8,18,13, 2 Technetium	44 **Ru** 2,8,18,15, 1 Ruthenium	45 **Rh** 2,8,18,16, 1 Rhodium	46 **Pd** 2,8,18, 18,0 Palladium	47 **Ag** 2,8,18, 18,1 Silver	48 **Cd** 2,8,18, 18,2 Cadmium	49 **In** 2,8,18, 18,3 Indium	50 **Sn** 2,8,18, 18,4 Tin	51 **Sb** 2,8,18, 18,5 Antimony	52 **Te** 2,8,18, 18,6 Tellurium	53 **I** 2,8,18, 18,7 Iodine	54 **Xe** 2,8,18, 18,8 Xenon
55 **Cs** 2,8,18,18, 8,1 Caesium	56 **Ba** 2,8,18,18, 8,2 Barium	57 **La** 2,8,18,18, 9,2 Lanthanum	72 **Hf** 2,8,18,32, 10,2 Hafnium	73 **Ta** 2,8,18, 32,11,2 Tantalum	74 **W** 2,8,18,32, 12,2 Tungsten	75 **Re** 2,8,18,32, 13,2 Rhenium	76 **Os** 2,8,18,32, 14,2 Osmium	77 **Ir** 2,8,18,32, 15,2 Iridium	78 **Pt** 2,8,18,32, 17,1 Platinum	79 **Au** 2,8,18, 32,18,1 Gold	80 **Hg** 2,8,18, 32,18,2 Mercury	81 **Tl** 2,8,18, 32,18,3 Thallium	82 **Pb** 2,8,18, 32,18,4 Lead	83 **Bi** 2,8,18, 32,18,5 Bismuth	84 **Po** 2,8,18, 32,18,6 Polonium	85 **At** 2,8,18, 32,18,7 Astatine	86 **Rn** 2,8,18, 32,18,8 Radon
87 **Fr** 2,8,18,32, 18,8,1 Francium	88 **Ra** 2,8,18,32, 18,8,2 Radium	89 **Ac** 2,8,18,32, 18,9,2 Actinium	104 **Rf** 2,8,18,32, 32,10,2 Rutherfordium	105 **Db** 2,8,18,32, 32,11,2 Dubnium	106 **Sg** 2,8,18,32, 32,12,2 Seaborgium	107 **Bh** 2,8,18,32, 32,13,2 Bohrium	108 **Hs** 2,8,18,32, 32,14,2 Hassium	109 **Mt** 2,8,18,32, 32,15,2 Meitnerium	110 **Ds** 2,8,18,32, 32,17,1 Darmstadtium	111 **Rg** 2,8,18,32, 32,18,1 Roentgenium	112 **Cn** 2,8,18,32, 32,18,2 Copernicium						

Transition Elements

Lanthanides

57 **La** 2,8,18, 18,9,2 Lanthanum	58 **Ce** 2,8,18, 20,8,2 Cerium	59 **Pr** 2,8,18,21, 8,2 Praseodymium	60 **Nd** 2,8,18,22, 8,2 Neodymium	61 **Pm** 2,8,18,23, 8,2 Promethium	62 **Sm** 2,8,18,24, 8,2 Samarium	63 **Eu** 2,8,18,25, 8,2 Europium	64 **Gd** 2,8,18,25, 9,2 Gadolinium	65 **Tb** 2,8,18,27, 8,2 Terbium	66 **Dy** 2,8,18,28, 8,2 Dysprosium	67 **Ho** 2,8,18,29, 8,2 Holmium	68 **Er** 2,8,18,30, 8,2 Erbium	69 **Tm** 2,8,18,31, 8,2 Thulium	70 **Yb** 2,8,18,32, 8,2 Ytterbium	71 **Lu** 2,8,18,32, 9,2 Lutetium

Actinides

89 **Ac** 2,8,18,32, 18,9,2 Actinium	90 **Th** 2,8,18,32, 18,10,2 Thorium	91 **Pa** 2,8,18,32, 20,9,2 Protactinium	92 **U** 2,8,18,32, 21,9,2 Uranium	93 **Np** 2,8,18,32, 22,9,2 Neptunium	94 **Pu** 2,8,18,32, 24,8,2 Plutonium	95 **Am** 2,8,18,32, 25,8,2 Americium	96 **Cm** 2,8,18,32, 25,9,2 Curium	97 **Bk** 2,8,18,32, 27,8,2 Berkelium	98 **Cf** 2,8,18,32, 28,8,2 Californium	99 **Es** 2,8,18,32, 29,8,2 Einsteinium	100 **Fm** 2,8,18,32, 30,8,2 Fermium	101 **Md** 2,8,18,32, 31,8,2 Mendelevium	102 **No** 2,8,18,32, 32,8,2 Nobelium	103 **Lr** 2,8,18,32, 32,9,2 Lawrencium

FOR OFFICIAL USE

N5

National
Qualifications
2014

Mark

X757/75/01

Physics
Section 1—Answer Grid and Section 2

THURSDAY, 22 MAY

9:00 AM – 11:00 AM

Fill in these boxes and read what is printed below.

Full name of centre

Town

Forename(s)

Surname

Number of seat

Date of birth

Day Month Year

D D M M Y Y

Scottish candidate number

Total marks — 110

SECTION 1 — 20 marks

Attempt ALL questions in this section.

Instructions for the completion of Section 1 are given on Page two.

SECTION 2 — 90 marks

Attempt ALL questions in this section.

Write your answers clearly in the spaces provided in this booklet. Additional space for answers and rough work is provided at the end of this booklet. If you use this space you must clearly identify the question number you are attempting. Any rough work must be written in this booklet. You should score through your rough work when you have written your final copy.

Use **blue** or **black** ink.

Reference may be made to the Data Sheet on Page two of the question paper X757/75/02 and to the Relationship Sheet X757/75/11.

Care should be taken to give an appropriate number of significant figures in the final answers to calculations.

Before leaving the examination room you must give this booklet to the Invigilator; if you do not, you may lose all the marks for this paper.

SECTION 1 — 20 marks

The questions for Section 1 are contained in the question paper X757/75/02.
Read these and record your answers on the answer grid on Page three opposite.
Do NOT use gel pens.

1. The answer to each question is **either** A, B, C, D or E. Decide what your answer is, then fill in the appropriate bubble (see sample question below).

2. There is **only one correct** answer to each question.

3. Any rough work must be written in the additional space for answers and rough work at the end of this booklet.

Sample Question

The energy unit measured by the electricity meter in your home is the:

A ampere

B kilowatt-hour

C watt

D coulomb

E volt.

The correct answer is **B**—kilowatt-hour. The answer **B** bubble has been clearly filled in (see below).

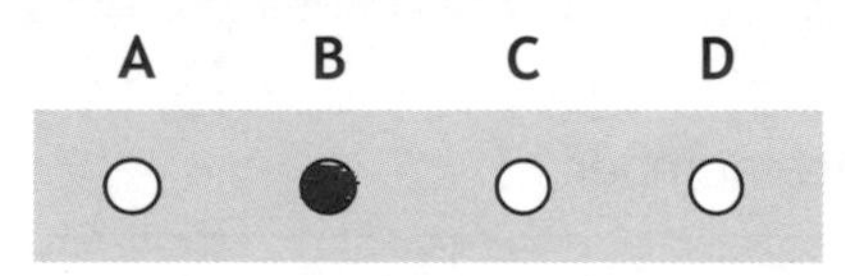

Changing an answer

If you decide to change your answer, cancel your first answer by putting a cross through it (see below) and fill in the answer you want. The answer below has been changed to **D**.

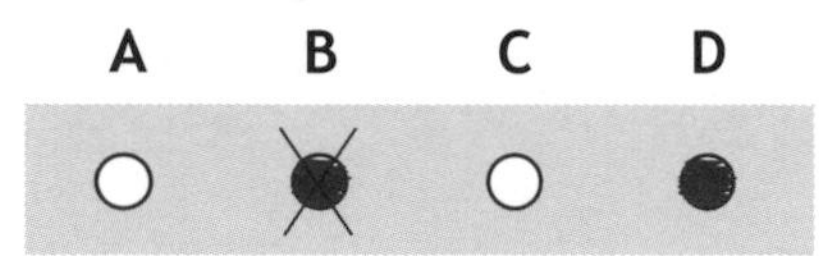

If you then decide to change back to an answer you have already scored out, put a tick (✓) to the **right** of the answer you want, as shown below:

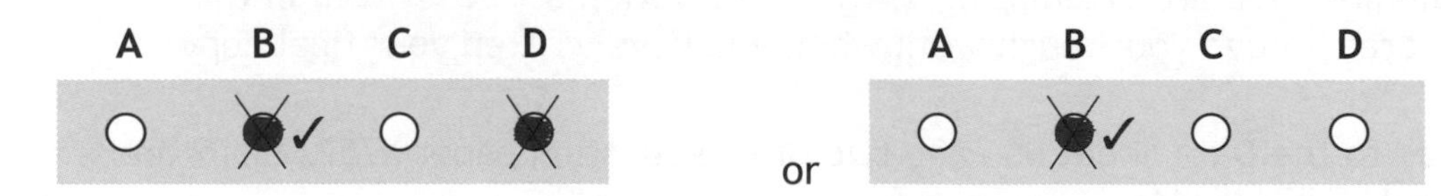

SECTION 1 — Answer Grid

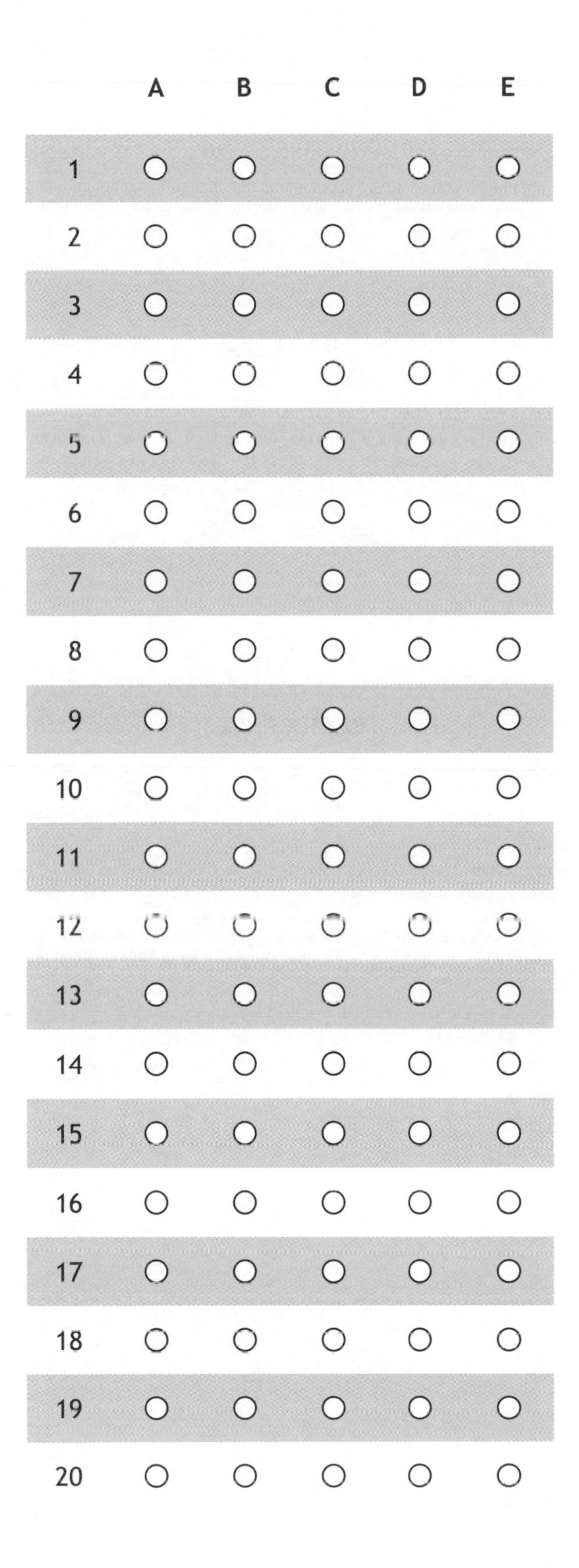

[Turn over

[BLANK PAGE]

DO NOT WRITE ON THIS PAGE

[Turn over for Question 1 on *Page six*

DO NOT WRITE ON THIS PAGE

SECTION 2 — 90 marks

Attempt ALL questions

1. A toy car contains an electric circuit which consists of a 12·0 V battery, an electric motor and two lamps.

The circuit diagram is shown.

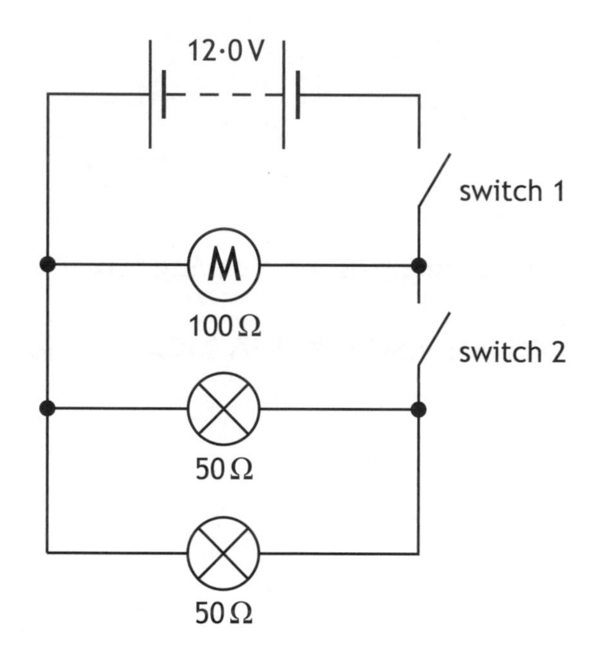

(a) Switch 1 is now closed.

Calculate the power dissipated in the motor when operating. **3**

Space for working and answer

MARKS | DO NOT WRITE IN THIS MARGIN

1. **(continued)**

(b) Switch 2 is now also closed.

(i) Calculate the total resistance of the motor and the two lamps. **3**

Space for working and answer

(ii) One of the lamps now develops a fault and stops working.

State the effect this has on the other lamp.

You **must** justify your answer. **2**

Total marks 8

[Turn over

MARKS | DO NOT WRITE IN THIS MARGIN

2. A thermistor is used as a temperature sensor in a circuit to monitor and control the temperature of water in a tank. Part of the circuit is shown.

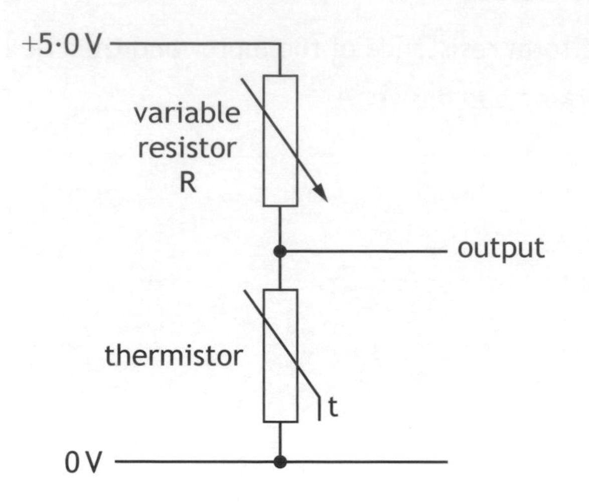

(a) (i) The variable resistor R is set at a resistance of 1050 Ω.

Calculate the resistance of the thermistor when the voltage across the thermistor is 2·0 V. **4**

Space for working and answer

MARKS | DO NOT WRITE IN THIS MARGIN

2. (a) (continued)

(ii) The graph shows how the resistance of the thermistor varies with temperature.

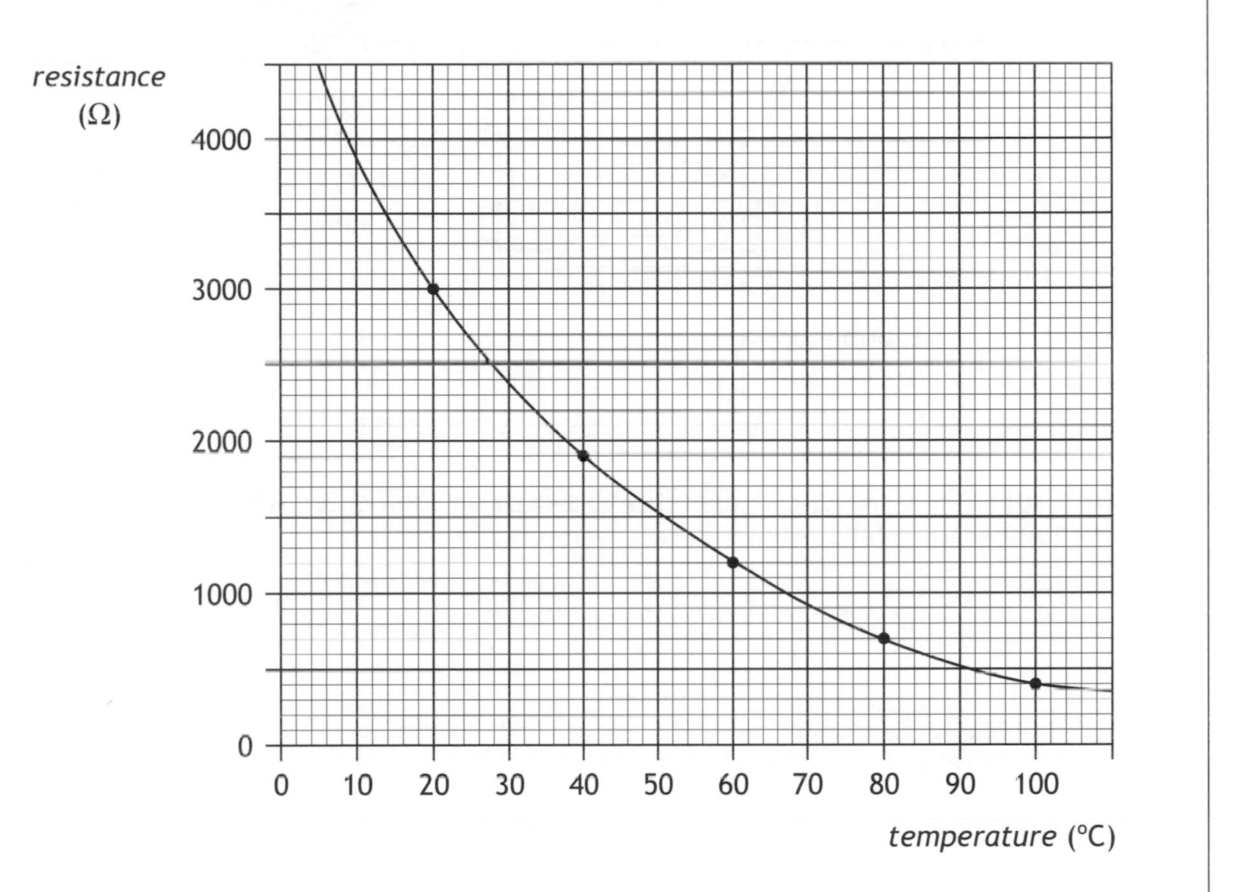

Use the graph to determine the temperature of the water when the voltage across the thermistor is 2·0 V. **1**

[Turn over

MARKS | DO NOT WRITE IN THIS MARGIN

2. (continued)

(b) The circuit is now connected to a switching circuit to operate a heater.

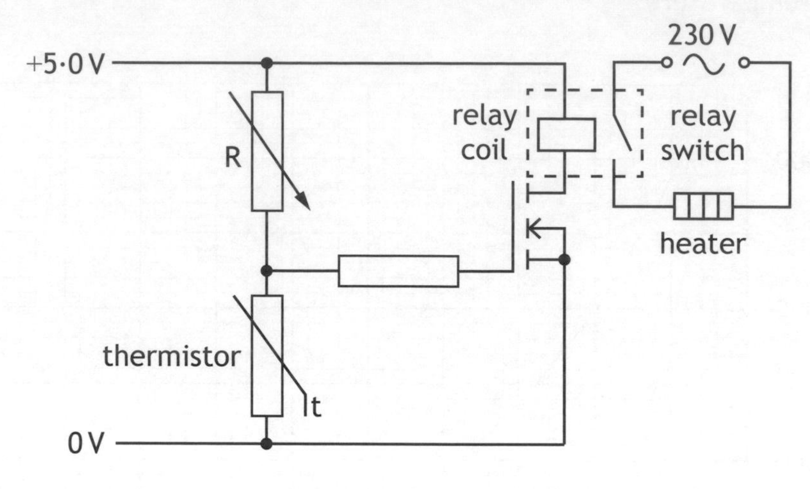

(i) Explain how the circuit operates to switch on the heater when the temperature falls below a certain value. **3**

(ii) The resistance of the variable resistor R is now increased.

What effect does this have on the temperature at which the heater is switched on?

You **must** justify your answer. **3**

Total marks 11

[Turn over for Question 3 on *Page twelve*

DO NOT WRITE ON THIS PAGE

MARKS | DO NOT WRITE IN THIS MARGIN

3. A student is investigating the specific heat capacity of three metal blocks X, Y and Z.

Each block has a mass of 1·0 kg.

A heater and thermometer are inserted into a block as shown.

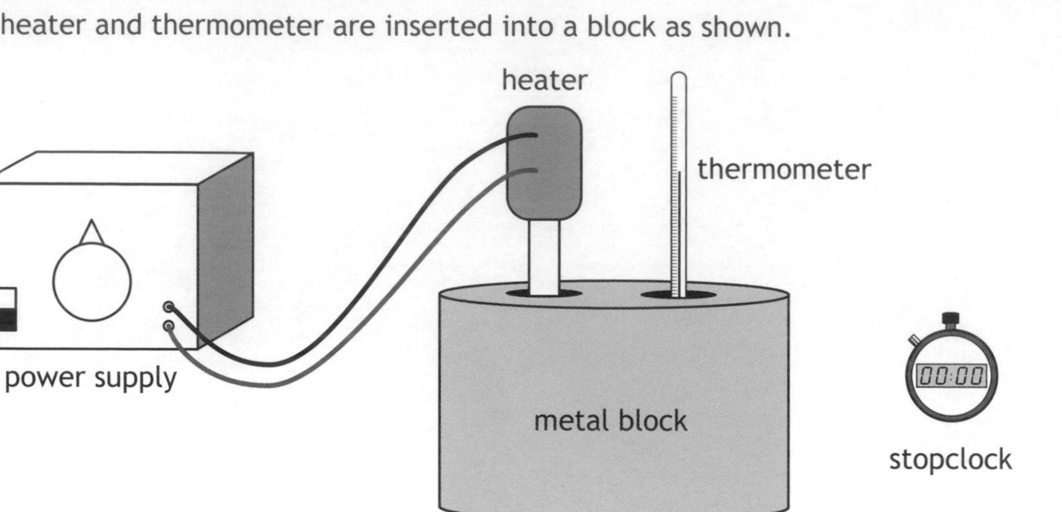

The heater has a power rating of 15 W.

The initial temperature of the block is measured.

The heater is switched on for 10 minutes and the final temperature of the block is recorded.

This procedure is repeated for the other two blocks.

The student's results are shown in the table.

Block	*Initial temperature* (°C)	*Final temperature* (°C)
X	15	25
Y	15	85
Z	15	34

(a) Show that the energy provided by the heater to each block is 9000 J. **2**

Space for working and answer

MARKS | DO NOT WRITE IN THIS MARGIN

3. (continued)

(b) (i) By referring to the results in the table, identify the block that has the greatest specific heat capacity. **1**

(ii) Calculate the specific heat capacity of the block identified in (b)(i). **3**

Space for working and answer

(c) Due to energy losses, the specific heat capacities calculated in this investigation are different from the accepted values.

The student decides to improve the set up in order to obtain a value closer to the accepted value for each block.

(i) Suggest a possible improvement that would reduce energy losses. **1**

(ii) State the effect that this improvement would have on the final temperature. **1**

Total marks 8

MARKS | DO NOT WRITE IN THIS MARGIN

4. A student, fishing from a pier, counts four waves passing the end of the pier in 20 seconds. The student estimates that the wavelength of the waves is 12 m.

Not to scale

(a) Calculate the speed of the water waves. **4**

Space for working and answer

MARKS | DO NOT WRITE IN THIS MARGIN

4. (continued)

(b) When looking down into the calm water behind the pier the student sees a fish.

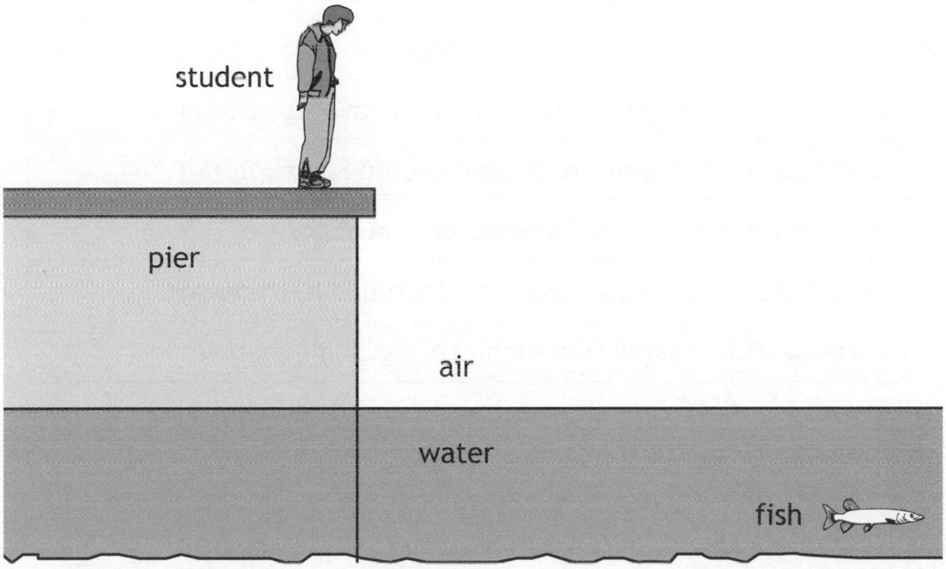

student

pier

air

water

fish

Complete the diagram to show the path of a ray of light from the fish to the student.

You should include the normal in your diagram. **3**

(An additional diagram, if required, can be found on *Page thirty-one*.)

Total marks 7

[Turn over

MARKS | DO NOT WRITE IN THIS MARGIN

5. The UV Index is an international standard measurement of the intensity of ultraviolet radiation from the Sun. Its purpose is to help people to effectively protect themselves from UV rays.

The UV index table is shown.

UV Index	*Description*
0 – 2	Low risk from the Sun's UV rays for the average person
3 – 5	Moderate risk of harm from unprotected Sun exposure
6 – 7	High risk of harm from unprotected Sun exposure
8 – 10	Very high risk of harm from unprotected Sun exposure
11+	Extreme risk of harm from unprotected Sun exposure

The UV index can be calculated using

$$UV\ index = \left[\begin{matrix} total\ effect\ of \\ UV\ radiation \end{matrix} \times \begin{matrix} elevation\ above \\ sea\ level\ adjustment \end{matrix} \times \begin{matrix} cloud \\ adjustment \end{matrix} \right] \div 25$$

The UV index is then rounded to the nearest whole number.

The tables below give information for elevation above sea level and cloud cover.

Elevation above sea level (km)	*Elevation above sea level adjustment*
1	1·06
2	1·12
3	1·18

Cloud cover	*Cloud adjustment*
Clear skies	1·00
Scattered clouds	0·89
Broken clouds	0·73
Overcast skies	0·31

5. (continued)

MARKS | DO NOT WRITE IN THIS MARGIN

(a) At a particular location the total effect of UV radiation is 280.

The elevation is 2 km above sea level with overcast skies.

Calculate the UV index value for this location. **2**

Space for working and answer

(b) Applying sunscreen to the skin is one method of protecting people from the Sun's harmful UV rays. UV radiation can be divided into three wavelength ranges, called UVA, UVB and UVC.

A manufacturer carries out some tests on experimental sunscreens P, Q and R to determine how effective they are at absorbing UV radiation. The test results are displayed in the graph.

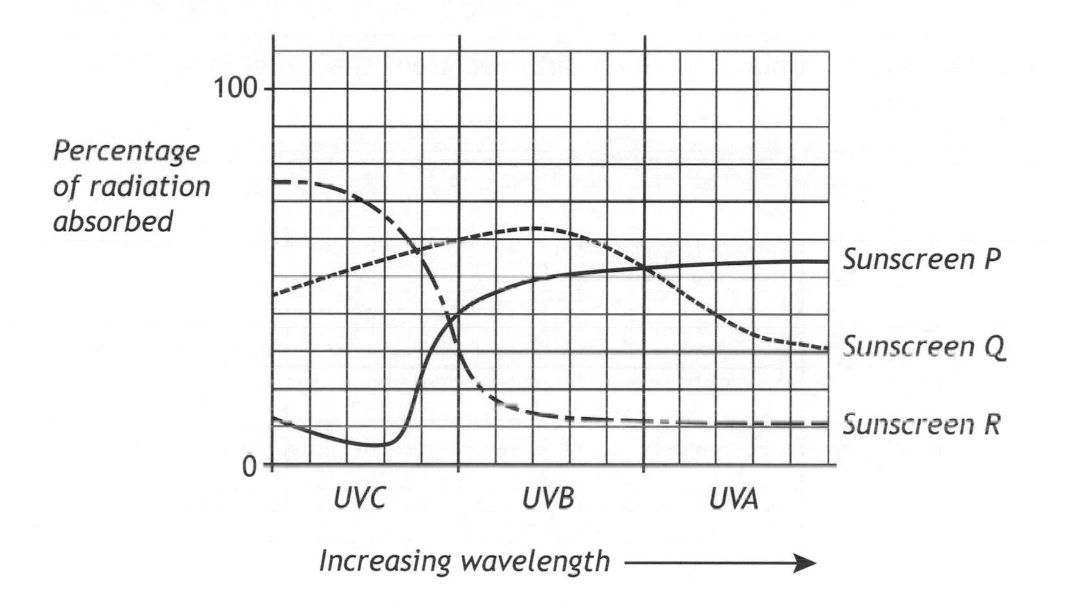

Using information from the graph, complete the following table. **2**

	UVA	*UVB*	*UVC*
Type of sunscreen that absorbs most of this radiation		Sunscreen Q	
Type of sunscreen that absorbs least of this radiation	Sunscreen R		

(c) State one useful application of UV radiation. **1**

Total marks 5

MARKS DO NOT WRITE IN THIS MARGIN

6. A technician carries out an experiment, using the apparatus shown, to determine the half-life of a radioactive source.

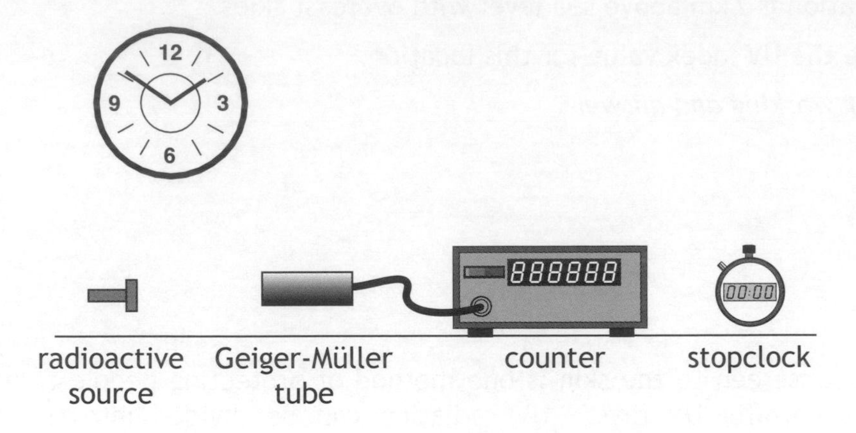

(a) State what is meant by the term *half-life*. **1**

(b) The technician displays the data obtained from the experiment in the graph below.

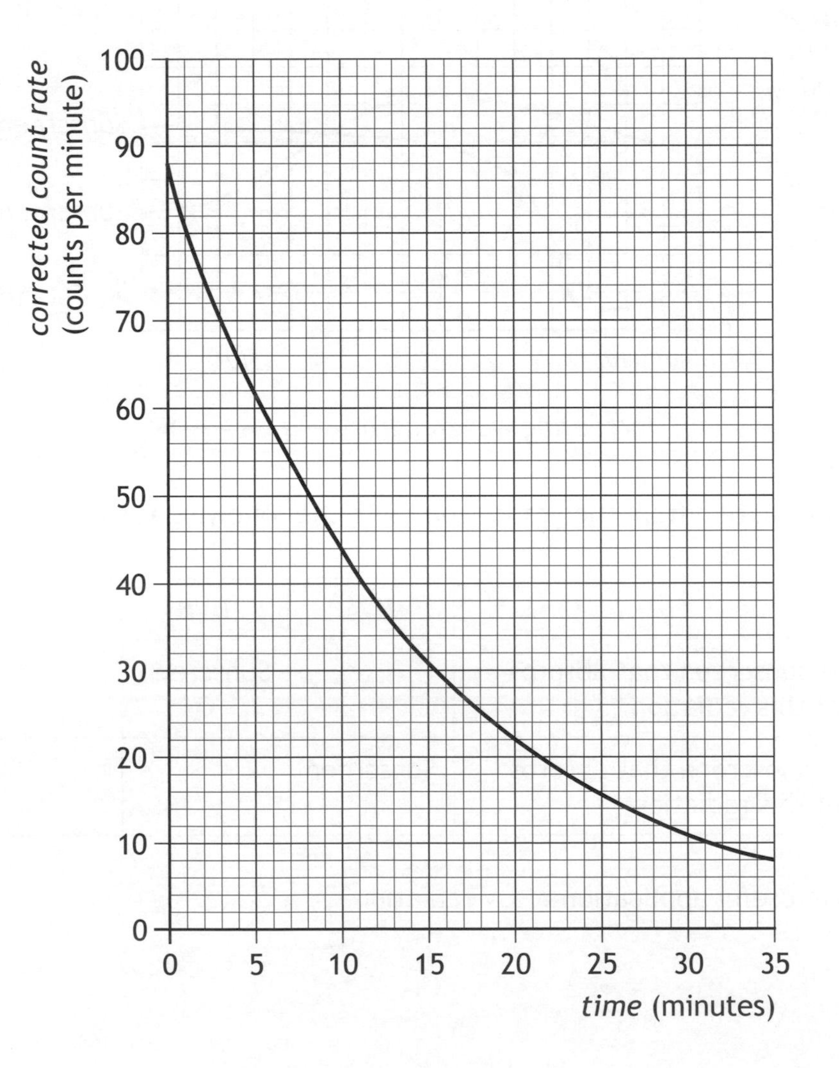

MARKS DO NOT WRITE IN THIS MARGIN

6. (b) (continued)

(i) Describe how the apparatus could be used to obtain the experimental data required to produce this graph. 3

(ii) Use information from the graph to determine the half-life of the radioactive source. 1

(iii) Determine the corrected count rate after 40 minutes. 2

Space for working and answer

Total marks 7

MARKS DO NOT WRITE IN THIS MARGIN

7. A fire engine on its way to an emergency is travelling along a main street. The siren on the fire engine is sounding.

A student standing in a nearby street cannot see the fire engine but can hear the siren.

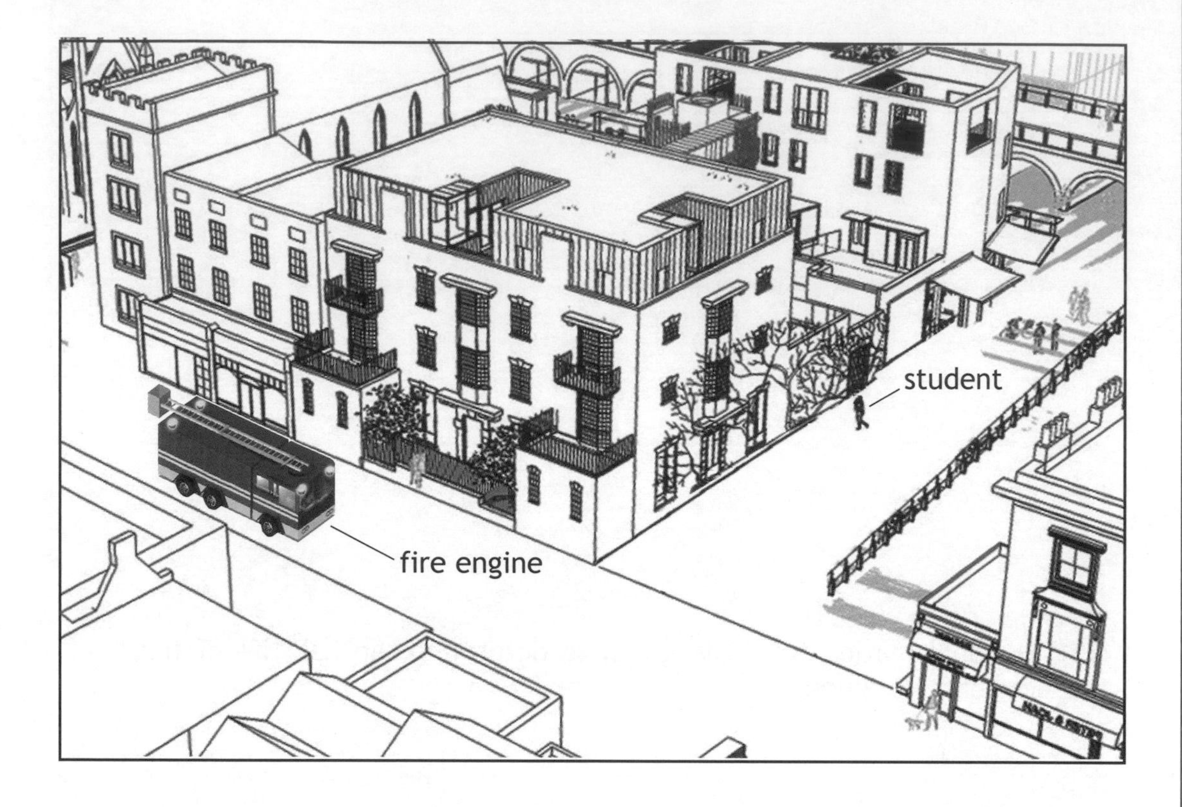

Use your knowledge of physics to comment on why the student can hear the siren even though the fire engine is not in view. **3**

MARKS DO NOT WRITE IN THIS MARGIN

8. An airport worker passes suitcases through an X-ray machine.

bibiphoto/shutterstock.com

(a) The worker has a mass of 80·0 kg and on a particular day absorbs 7·2 mJ of energy from the X-ray machine.

(i) Calculate the absorbed dose received by the worker. **3**

Space for working and answer

(ii) Calculate the equivalent dose received by the worker. **3**

Space for working and answer

MARKS | DO NOT WRITE IN THIS MARGIN

8. (continued)

(b) X-rays can cause ionisation.

Explain what is meant by *ionisation*. 1

Total marks 7

MARKS DO NOT WRITE IN THIS MARGIN

9. A communications satellite is used to transmit live television broadcasts from the UK to Canada.

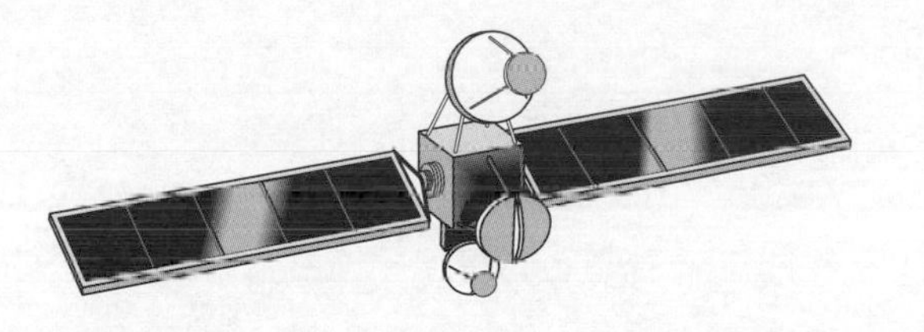

A student states that, to allow the live television broadcasts to be received in Canada, it is important that the satellite does not move.

Use your knowledge of physics to comment on this statement. **3**

[Turn over

MARKS | DO NOT WRITE IN THIS MARGIN

10. In a rowing event a boat moves off in a straight line.

A graph of the boat's motion is shown.

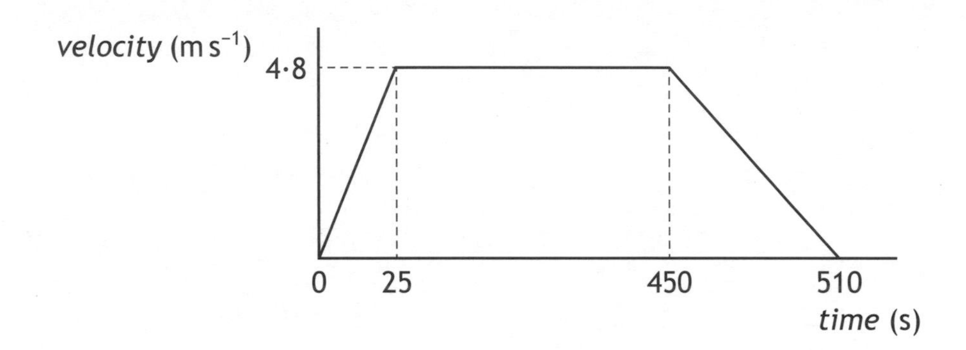

(a) (i) Calculate the acceleration of the boat during the first 25 s. **3**

Space for working and answer

(ii) Describe the motion of the boat between 25 s and 450 s. **1**

MARKS DO NOT WRITE IN THIS MARGIN

10. (a) (continued)

(iii) Draw a diagram showing the horizontal forces acting on the boat between 25 s and 450 s.

You **must** name these forces and show their directions. **2**

(b) The boat comes to rest after 510 s.

(i) Calculate the total distance travelled by the boat. **3**

Space for working and answer

(ii) Calculate the average velocity of the boat.

A direction is not required. **3**

Space for working and answer

Total marks 12

MARKS | DO NOT WRITE IN THIS MARGIN

11. A helicopter is used to take tourists on sightseeing flights.

Information about the helicopter is shown in the table.

weight of empty helicopter	13 500 N
maximum take-off weight	24 000 N
cruising speed	$67\,m\,s^{-1}$
maximum speed	$80\,m\,s^{-1}$
maximum range	610 km

(a) The pilot and passengers are weighed before they board the helicopter.

Explain the reason for this. **1**

(b) Six passengers and the pilot with a combined weight of 6125 N board the helicopter.

Determine the minimum upward force required by the helicopter at take-off. **1**

Space for working and answer

MARKS | DO NOT WRITE IN THIS MARGIN

11. (continued)

(c) The helicopter travels 201 km at its cruising speed.

Calculate the time taken to travel this distance. **3**

Space for working and answer

Total marks 5

[Turn over

MARKS | DO NOT WRITE IN THIS MARGIN

12. A student is investigating the motion of water rockets. The water rocket is made from an upturned plastic bottle containing some water. Air is pumped into the bottle. When the pressure of the air is great enough the plastic bottle is launched upwards.

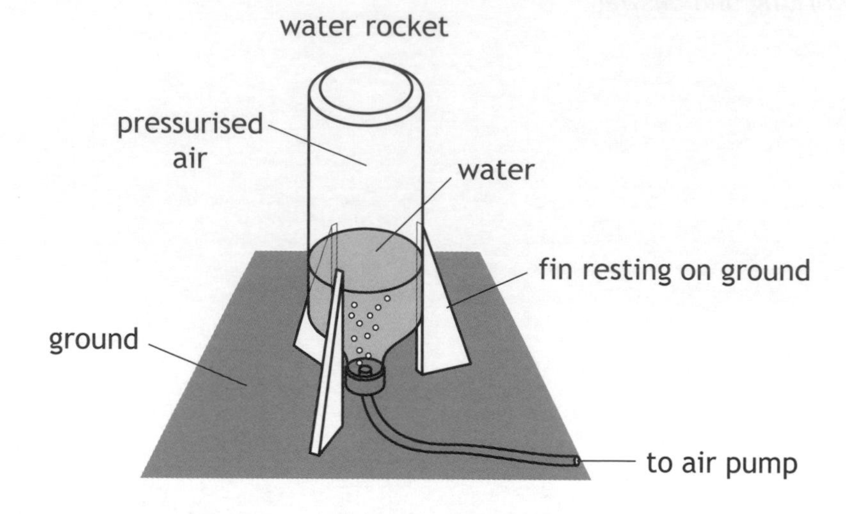

The mass of the rocket before launch is 0·94 kg.

(a) Calculate the weight of the water rocket. **3**

Space for working and answer

(b) Before launch, the water rocket rests on three fins on the ground.

The area of each fin in contact with the ground is $2{\cdot}0 \times 10^{-4}\,m^2$.

Calculate the total pressure exerted on the ground by the fins. **4**

Space for working and answer

MARKS DO NOT WRITE IN THIS MARGIN

12. (continued)

(c) Use Newton's Third Law to explain how the rocket launches. 1

(d) At launch, the initial upward thrust on the rocket is 370 N.

Calculate the initial acceleration of the rocket. 4

Space for working and answer

(e) The student launches the rocket a second time.

For this launch, the student adds a greater volume of water than before.

The same initial upward thrust acts on the rocket but it fails to reach the same height.

Explain why the rocket fails to reach the same height. 2

Total marks 14

[END OF QUESTION PAPER]

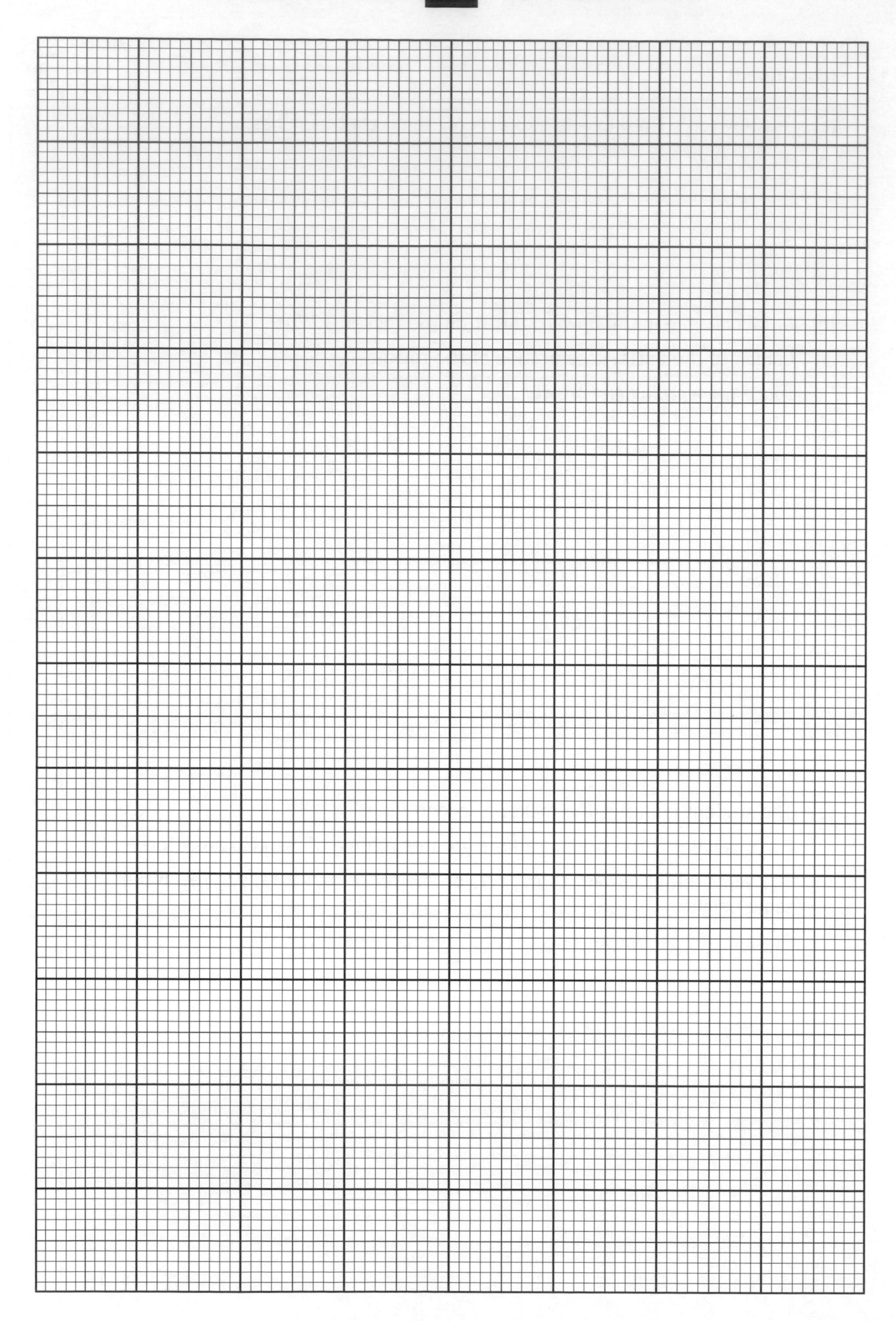

MARKS DO NOT WRITE IN THIS MARGIN

ADDITIONAL SPACE FOR ANSWERS AND ROUGH WORK

Additional diagram for Question 4 (b)

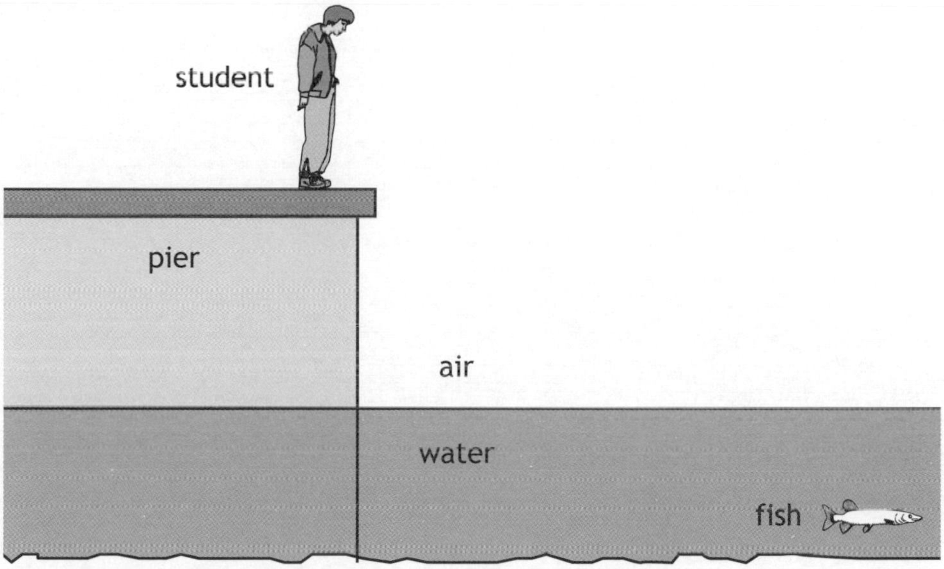

student

pier

air

water

fish

MARKS | DO NOT WRITE IN THIS MARGIN

ADDITIONAL SPACE FOR ANSWERS AND ROUGH WORK

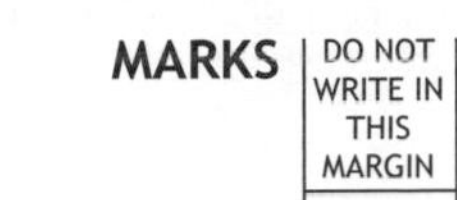

ADDITIONAL SPACE FOR ANSWERS AND ROUGH WORK

[BLANK PAGE]

DO NOT WRITE ON THIS PAGE

NATIONAL 5
2015

SQA

HODDER
GIBSON
LEARN MORE

National
Qualifications
2015

X757/75/02

Physics
Section 1—Questions

TUESDAY, 5 MAY

9:00 AM – 11:00 AM

Instructions for the completion of Section 1 are given on *Page two* of your question and answer booklet X757/75/01.

Record your answers on the answer grid on *Page three* of your question and answer booklet.

Reference may be made to the Data Sheet on *Page two* of this booklet and to the Relationship Sheet X757/75/11.

Before leaving the examination room you must give your question and answer booklet to the Invigilator; if you do not, you may lose all the marks for this paper.

DATA SHEET

Speed of light in materials

Material	*Speed in* $m\,s^{-1}$
Air	$3{\cdot}0 \times 10^8$
Carbon dioxide	$3{\cdot}0 \times 10^8$
Diamond	$1{\cdot}2 \times 10^8$
Glass	$2{\cdot}0 \times 10^8$
Glycerol	$2{\cdot}1 \times 10^8$
Water	$2{\cdot}3 \times 10^8$

Gravitational field strengths

	Gravitational field strength on the surface in $N\,kg^{-1}$
Earth	9·8
Jupiter	23
Mars	3·7
Mercury	3·7
Moon	1·6
Neptune	11
Saturn	9·0
Sun	270
Uranus	8·7
Venus	8·9

Specific latent heat of fusion of materials

Material	*Specific latent heat of fusion in* $J\,kg^{-1}$
Alcohol	$0{\cdot}99 \times 10^5$
Aluminium	$3{\cdot}95 \times 10^5$
Carbon Dioxide	$1{\cdot}80 \times 10^5$
Copper	$2{\cdot}05 \times 10^5$
Iron	$2{\cdot}67 \times 10^5$
Lead	$0{\cdot}25 \times 10^5$
Water	$3{\cdot}34 \times 10^5$

Specific latent heat of vaporisation of materials

Material	*Specific latent heat of vaporisation in* $J\,kg^{-1}$
Alcohol	$11{\cdot}2 \times 10^5$
Carbon Dioxide	$3{\cdot}77 \times 10^5$
Glycerol	$8{\cdot}30 \times 10^5$
Turpentine	$2{\cdot}90 \times 10^5$
Water	$22{\cdot}6 \times 10^5$

Speed of sound in materials

Material	*Speed in* $m\,s^{-1}$
Aluminium	5200
Air	340
Bone	4100
Carbon dioxide	270
Glycerol	1900
Muscle	1600
Steel	5200
Tissue	1500
Water	1500

Specific heat capacity of materials

Material	*Specific heat capacity in* $J\,kg^{-1}\,°C^{-1}$
Alcohol	2350
Aluminium	902
Copper	386
Glass	500
Ice	2100
Iron	480
Lead	128
Oil	2130
Water	4180

Melting and boiling points of materials

Material	*Melting point in* °C	*Boiling point in* °C
Alcohol	−98	65
Aluminium	660	2470
Copper	1077	2567
Glycerol	18	290
Lead	328	1737
Iron	1537	2737

Radiation weighting factors

Type of radiation	*Radiation weighting factor*
alpha	20
beta	1
fast neutrons	10
gamma	1
slow neutrons	3
X-rays	1

SECTION 1

Attempt ALL questions

1. Two circuits are set up as shown.

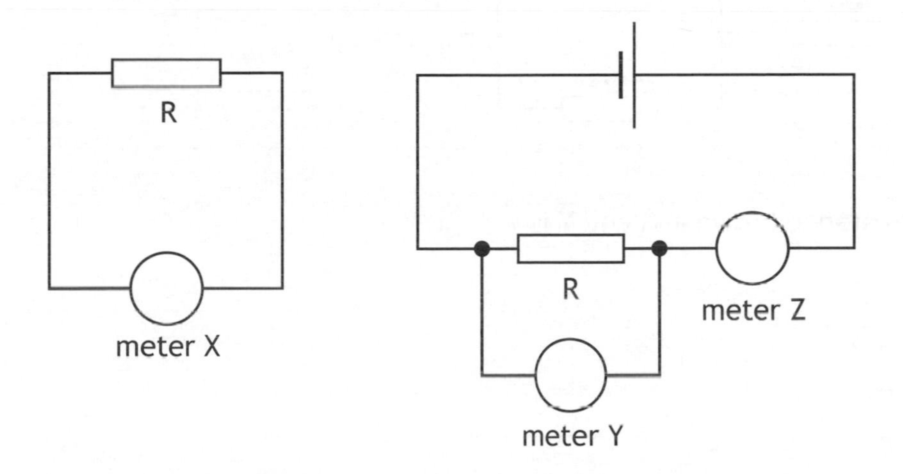

Both circuits are used to determine the resistance of resistor R.

Which row in the table identifies meter X, meter Y and meter Z?

	meter X	*meter Y*	*meter Z*
A	ohmmeter	voltmeter	ammeter
B	ohmmeter	ammeter	voltmeter
C	voltmeter	ammeter	ohmmeter
D	ammeter	voltmeter	ohmmeter
E	voltmeter	ohmmeter	ammeter

2. Which of the following statements is/are correct?

I The voltage of a battery is the number of joules of energy it gives to each coulomb of charge.

II A battery only has a voltage when it is connected in a complete circuit.

III Electrons are free to move within an insulator.

A I only

B II only

C III only

D II and III only

E I, II and III

[Turn over

3. A circuit is set up as shown.

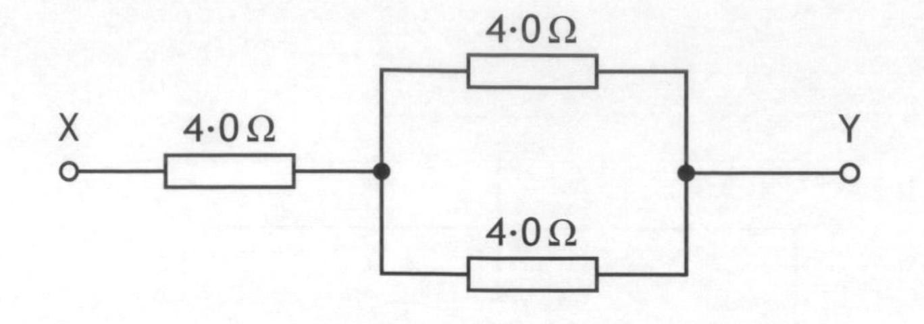

The resistance between X and Y is

A 1·3 Ω

B 4·5 Ω

C 6·0 Ω

D 8·0 Ω

E 12 Ω.

4. The rating plate on an electrical appliance is shown.

230 V~
50 Hz
920 W
model: HD 1055

The resistance of this appliance is

A 0·017 Ω

B 0·25 Ω

C 4·0 Ω

D 18·4 Ω

E 57·5 Ω.

5. A syringe containing air is sealed at one end as shown.

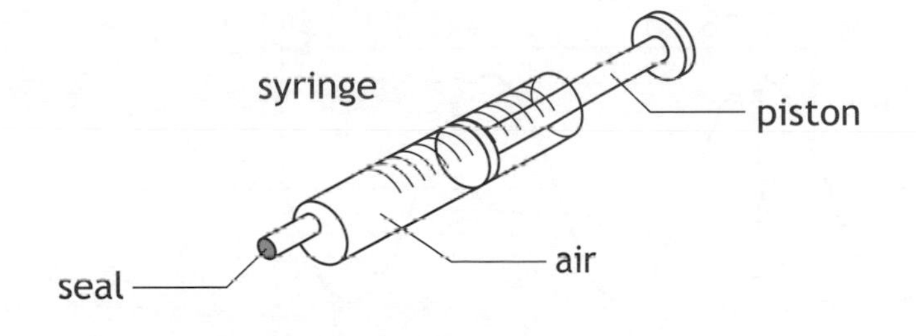

The piston is pushed in slowly.

There is no change in temperature of the air inside the syringe.

Which of the following statements describes and explains the change in pressure of the air in the syringe?

A The pressure increases because the air particles have more kinetic energy.

B The pressure increases because the air particles hit the sides of the syringe more frequently.

C The pressure increases because the air particles hit the sides of the syringe less frequently.

D The pressure decreases because the air particles hit the sides of the syringe with less force.

E The pressure decreases because the air particles have less kinetic energy.

6. The pressure of a fixed mass of gas is 150 kPa at a temperature of 27 °C.

The temperature of the gas is now increased to 47 °C.

The volume of the gas remains constant.

The pressure of the gas is now

A 86 kPa

B 141 kPa

C 150 kPa

D 160 kPa

E 261 kPa.

[Turn over

7. The diagram represents a water wave.

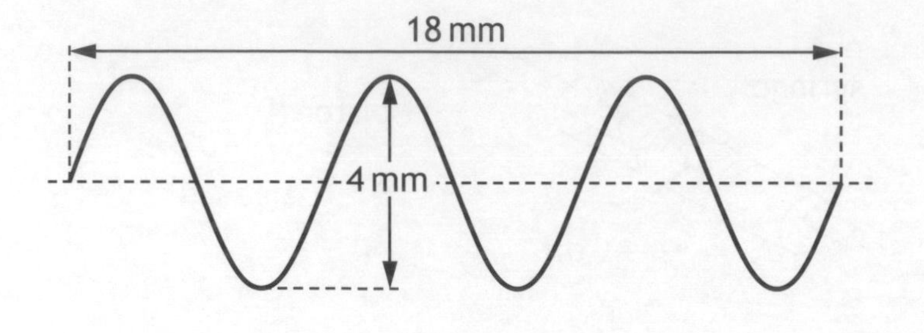

The wavelength of the water wave is

A 2 mm

B 3 mm

C 4 mm

D 6 mm

E 18 mm.

8. A student makes the following statements about different types of electromagnetic waves.

I Light waves are transverse waves.

II Radio waves travel at $340\,m\,s^{-1}$ through air.

III Ultraviolet waves have a longer wavelength than infrared waves.

Which of these statements is/are correct?

A I only

B I and II only

C I and III only

D II and III only

E I, II and III

9. Alpha radiation ionises an atom.

Which statement describes what happens to the atom?

A The atom splits in half.

B The atom releases a neutron.

C The atom becomes positively charged.

D The atom gives out gamma radiation.

E The atom releases heat.

10. A sample of tissue is irradiated using a radioactive source.

A student makes the following statements.

The equivalent dose received by the tissue is

I reduced by shielding the tissue with a lead screen

II increased as the distance from the source to the tissue is increased

III increased by increasing the time of exposure of the tissue to the radiation.

Which of the statements is/are correct?

A I only

B II only

C I and II only

D II and III only

E I and III only

11. A sample of tissue receives an absorbed dose of 16 μGy from alpha particles.

The radiation weighting factor for alpha particles is 20.

The equivalent dose received by the sample is

A 0·80 μSv

B 1·25 μSv

C 4 μSv

D 36 μSv

E 320 μSv.

12. For a particular radioactive source, 240 atoms decay in 1 minute.

The activity of this source is

A 4 Bq

B 180 Bq

C 240 Bq

D 300 Bq

E 14 400 Bq.

[Turn over

13. The letters **X**, **Y** and **Z** represent missing words from the following passage.

During a nuclear ......X...... reaction two nuclei of smaller mass number combine to produce a nucleus of larger mass number. During a nuclear ......Y...... reaction a nucleus of larger mass number splits into two nuclei of smaller mass number. Both of these reactions are important because these processes can release ......Z...... .

Which row in the table shows the missing words?

	X	*Y*	*Z*
A	fusion	fission	electrons
B	fission	fusion	energy
C	fusion	fission	protons
D	fission	fusion	protons
E	fusion	fission	energy

14. Which of the following quantities is fully described by its magnitude?

A Force

B Displacement

C Energy

D Velocity

E Acceleration

15. The table shows the velocities of three objects X, Y and Z over a period of 3 seconds. Each object is moving in a straight line.

Time (s)	0	1	2	3
Velocity of X ($m\,s^{-1}$)	2	4	6	8
Velocity of Y ($m\,s^{-1}$)	0	1	2	3
Velocity of Z ($m\,s^{-1}$)	0	2	5	9

Which of the following statements is/are correct?

I X moves with constant velocity.

II Y moves with constant acceleration.

III Z moves with constant acceleration.

A I only

B II only

C I and II only

D I and III only

E II and III only

16. A car of mass 1200 kg is travelling along a straight level road at a constant speed of $20\,m\,s^{-1}$.

The driving force on the car is 2500 N. The frictional force on the car is 2500 N.

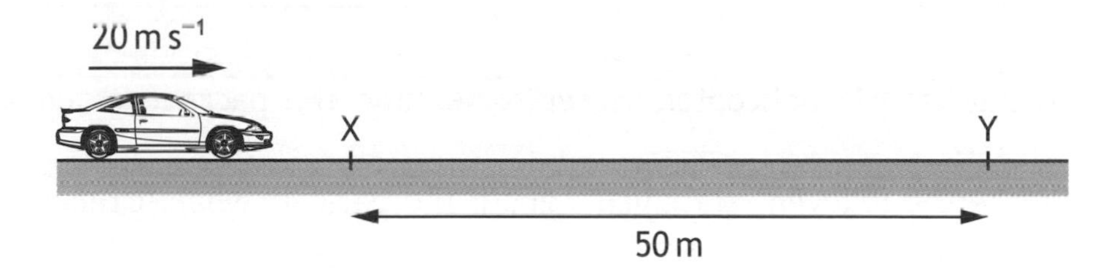

The work done moving the car between point X and point Y is

A 0 J

B 11 800 J

C 125 000 J

D 240 000 J

E 250 000 J.

[Turn over

17. A person sits on a chair which rests on the Earth. The person exerts a downward force on the chair.

Which of the following is the reaction to this force?

A The force of the chair on the person

B The force of the person on the chair

C The force of the Earth on the person

D The force of the chair on the Earth

E The force of the person on the Earth

18. A package falls vertically from a helicopter. After some time the package reaches its terminal velocity.

A group of students make the following statements about the package when it reaches its terminal velocity.

I The weight of the package is less than the air resistance acting on the package.

II The forces acting on the package are balanced.

III The package is accelerating towards the ground at $9 \cdot 8\,m\,s^{-2}$.

Which of these statements is/are correct?

A I only

B II only

C III only

D I and III only

E II and III only

19. The distance from the Sun to Proxima Centauri is 4·3 light years.

This distance is equivalent to

A $1 \cdot 4 \times 10^{8}$ m

B $1 \cdot 6 \times 10^{14}$ m

C $6 \cdot 8 \times 10^{14}$ m

D $9 \cdot 5 \times 10^{15}$ m

E $4 \cdot 1 \times 10^{16}$ m.

20. Light from a star is split into a line spectrum of different colours. The line spectrum from the star is shown, along with the line spectra of the elements calcium, helium, hydrogen and sodium.

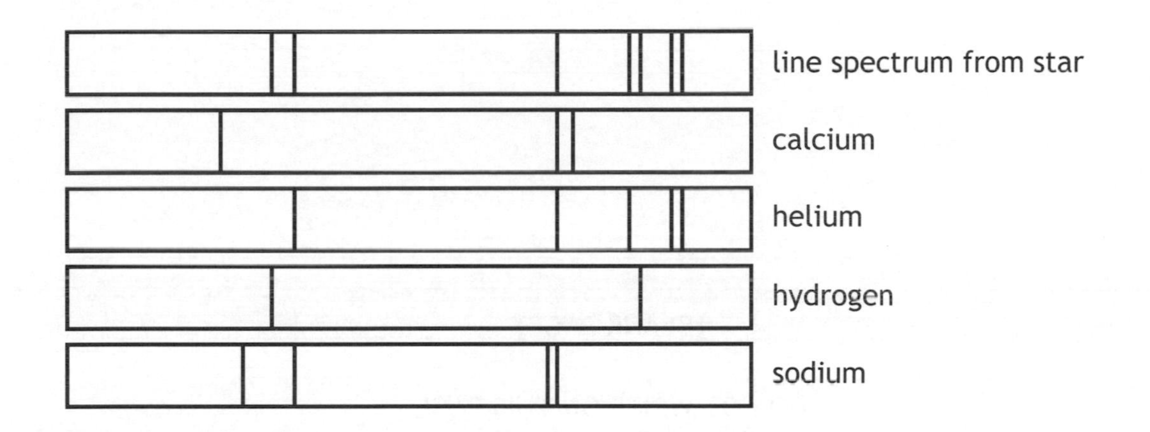

The elements present in this star are

A sodium and calcium

B calcium and helium

C hydrogen and sodium

D helium and hydrogen

E calcium, sodium and hydrogen.

[END OF SECTION 1. NOW ATTEMPT THE QUESTIONS IN SECTION 2 OF YOUR QUESTION AND ANSWER BOOKLET]

[BLANK PAGE]

DO NOT WRITE ON THIS PAGE

National
Qualifications
2015

X757/75/11

Physics
Relationships Sheet

TUESDAY, 5 MAY

9:00 AM – 11:00 AM

$E_p = mgh$

$E_k = \frac{1}{2}mv^2$

$Q = It$

$V = IR$

$R_T = R_1 + R_2 + \ldots$

$\frac{1}{R_T} = \frac{1}{R_1} + \frac{1}{R_2} + \ldots$

$V_2 = \left(\frac{R_2}{R_1 + R_2}\right)V_s$

$\frac{V_1}{V_2} = \frac{R_1}{R_2}$

$P = \frac{E}{t}$

$P = IV$

$P = I^2R$

$P = \frac{V^2}{R}$

$E_h = cm\Delta T$

$p = \frac{F}{A}$

$\frac{pV}{T} = \text{constant}$

$p_1V_1 = p_2V_2$

$\frac{p_1}{T_1} = \frac{p_2}{T_2}$

$\frac{V_1}{T_1} = \frac{V_2}{T_2}$

$d = vt$

$v = f\lambda$

$T = \frac{1}{f}$

$A = \frac{N}{t}$

$D = \frac{E}{m}$

$H = Dw_R$

$\dot{H} = \frac{H}{t}$

$s = vt$

$d = \bar{v}t$

$s = \bar{v}t$

$a = \frac{v-u}{t}$

$W = mg$

$F = ma$

$E_w = Fd$

$E_h = ml$

Additional Relationships

Circle

circumference = $2\pi r$

area = πr^2

Sphere

area = $4\pi r^2$

volume = $\frac{4}{3}\pi r^3$

Trigonometry

$\sin\theta = \frac{\text{opposite}}{\text{hypotenuse}}$

$\cos\theta = \frac{\text{adjacent}}{\text{hypotenuse}}$

$\tan\theta = \frac{\text{opposite}}{\text{adjacent}}$

$\sin^2\theta + \cos^2\theta = 1$

Electron Arrangements of Elements

Key

Atomic number
Symbol
Electron arrangement
Name

Transition Elements

Group 1 (1)	Group 2 (2)	(3)	(4)	(5)	(6)	(7)	(8)	(9)	(10)	(11)	(12)	Group 3 (13)	Group 4 (14)	Group 5 (15)	Group 6 (16)	Group 7 (17)	Group 0 (18)
1 **H** 1 Hydrogen																	2 **He** 2 Helium
3 **Li** 2,1 Lithium	4 **Be** 2,2 Beryllium											5 **B** 2,3 Boron	6 **C** 2,4 Carbon	7 **N** 2,5 Nitrogen	8 **O** 2,6 Oxygen	9 **F** 2,7 Fluorine	10 **Ne** 2,8 Neon
11 **Na** 2,8,1 Sodium	12 **Mg** 2,8,2 Magnesium											13 **Al** 2,8,3 Aluminium	14 **Si** 2,8,4 Silicon	15 **P** 2,8,5 Phosphorus	16 **S** 2,8,6 Sulfur	17 **Cl** 2,8,7 Chlorine	18 **Ar** 2,8,8 Argon
19 **K** 2,8,8,1 Potassium	20 **Ca** 2,8,8,2 Calcium	21 **Sc** 2,8,9,2 Scandium	22 **Ti** 2,8,10,2 Titanium	23 **V** 2,8,11,2 Vanadium	24 **Cr** 2,8,13,1 Chromium	25 **Mn** 2,8,13,2 Manganese	26 **Fe** 2,8,14,2 Iron	27 **Co** 2,8,15,2 Cobalt	28 **Ni** 2,8,16,2 Nickel	29 **Cu** 2,8,18,1 Copper	30 **Zn** 2,8,18,2 Zinc	31 **Ga** 2,8,18,3 Gallium	32 **Ge** 2,8,18,4 Germanium	33 **As** 2,8,18,5 Arsenic	34 **Se** 2,8,18,6 Selenium	35 **Br** 2,8,18,7 Bromine	36 **Kr** 2,8,18,8 Krypton
37 **Rb** 2,8,18,8,1 Rubidium	38 **Sr** 2,8,18,8,2 Strontium	39 **Y** 2,8,18,9,2 Yttrium	40 **Zr** 2,8,18,10,2 Zirconium	41 **Nb** 2,8,18,12,1 Niobium	42 **Mo** 2,8,18,13,1 Molybdenum	43 **Tc** 2,8,18,13,2 Technetium	44 **Ru** 2,8,18,15,1 Ruthenium	45 **Rh** 2,8,18,16,1 Rhodium	46 **Pd** 2,8,18,18,0 Palladium	47 **Ag** 2,8,18,18,1 Silver	48 **Cd** 2,8,18,18,2 Cadmium	49 **In** 2,8,18,18,3 Indium	50 **Sn** 2,8,18,18,4 Tin	51 **Sb** 2,8,18,18,5 Antimony	52 **Te** 2,8,18,18,6 Tellurium	53 **I** 2,8,18,18,7 Iodine	54 **Xe** 2,8,18,18,8 Xenon
55 **Cs** 2,8,18,18,8,1 Caesium	56 **Ba** 2,8,18,18,8,2 Barium	57 **La** 2,8,18,18,9,2 Lanthanum	72 **Hf** 2,8,18,32,10,2 Hafnium	73 **Ta** 2,8,18,32,11,2 Tantalum	74 **W** 2,8,18,32,12,2 Tungsten	75 **Re** 2,8,18,32,13,2 Rhenium	76 **Os** 2,8,18,32,14,2 Osmium	77 **Ir** 2,8,18,32,15,2 Iridium	78 **Pt** 2,8,18,32,17,1 Platinum	79 **Au** 2,8,18,32,18,1 Gold	80 **Hg** 2,8,18,32,18,2 Mercury	81 **Tl** 2,8,18,32,18,3 Thallium	82 **Pb** 2,8,18,32,18,4 Lead	83 **Bi** 2,8,18,32,18,5 Bismuth	84 **Po** 2,8,18,32,18,6 Polonium	85 **At** 2,8,18,32,18,7 Astatine	86 **Rn** 2,8,18,32,18,8 Radon
87 **Fr** 2,8,18,32,18,8,1 Francium	88 **Ra** 2,8,18,32,18,8,2 Radium	89 **Ac** 2,8,18,32,18,9,2 Actinium	104 **Rf** 2,8,18,32,32,10,2 Rutherfordium	105 **Db** 2,8,18,32,32,11,2 Dubnium	106 **Sg** 2,8,18,32,32,12,2 Seaborgium	107 **Bh** 2,8,18,32,32,13,2 Bohrium	108 **Hs** 2,8,18,32,32,14,2 Hassium	109 **Mt** 2,8,18,32,32,15,2 Meitnerium	110 **Ds** 2,8,18,32,32,17,1 Darmstadtium	111 **Rg** 2,8,18,32,32,18,1 Roentgenium	112 **Cn** 2,8,18,32,32,18,2 Copernicium						

Lanthanides	57 **La** 2,8,18,18,9,2 Lanthanum	58 **Ce** 2,8,18,20,8,2 Cerium	59 **Pr** 2,8,18,21,8,2 Praseodymium	60 **Nd** 2,8,18,22,8,2 Neodymium	61 **Pm** 2,8,18,23,8,2 Promethium	62 **Sm** 2,8,18,24,8,2 Samarium	63 **Eu** 2,8,18,25,8,2 Europium	64 **Gd** 2,8,18,25,9,2 Gadolinium	65 **Tb** 2,8,18,27,8,2 Terbium	66 **Dy** 2,8,18,28,8,2 Dysprosium	67 **Ho** 2,8,18,29,8,2 Holmium	68 **Er** 2,8,18,30,8,2 Erbium	69 **Tm** 2,8,18,31,8,2 Thulium	70 **Yb** 2,8,18,32,8,2 Ytterbium	71 **Lu** 2,8,18,32,9,2 Lutetium
Actinides	89 **Ac** 2,8,18,32,18,9,2 Actinium	90 **Th** 2,8,18,32,18,10,2 Thorium	91 **Pa** 2,8,18,32,20,9,2 Protactinium	92 **U** 2,8,18,32,21,9,2 Uranium	93 **Np** 2,8,18,32,22,9,2 Neptunium	94 **Pu** 2,8,18,32,24,8,2 Plutonium	95 **Am** 2,8,18,32,25,8,2 Americium	96 **Cm** 2,8,18,32,25,9,2 Curium	97 **Bk** 2,8,18,32,27,8,2 Berkelium	98 **Cf** 2,8,18,32,28,8,2 Californium	99 **Es** 2,8,18,32,29,8,2 Einsteinium	100 **Fm** 2,8,18,32,30,8,2 Fermium	101 **Md** 2,8,18,32,31,8,2 Mendelevium	102 **No** 2,8,18,32,32,8,2 Nobelium	103 **Lr** 2,8,18,32,32,9,2 Lawrencium

FOR OFFICIAL USE

N5

National
Qualifications
2015

Mark

X757/75/01

Physics
Section 1—Answer Grid and Section 2

TUESDAY, 5 MAY

9:00 AM – 11:00 AM

Fill in these boxes and read what is printed below.

Full name of centre

Town

Forename(s)

Surname

Number of seat

Date of birth

Day Month Year

Scottish candidate number

Total marks — 110

SECTION 1 — 20 marks

Attempt ALL questions.

Instructions for the completion of Section 1 are given on *Page two*.

SECTION 2 — 90 marks

Attempt ALL questions.

Reference may be made to the Data Sheet on *Page two* of the question paper X757/75/02 and to the Relationship Sheet X757/75/11.

Care should be taken to give an appropriate number of significant figures in the final answers to calculations.

Write your answers clearly in the spaces provided in this booklet. Additional space for answers and rough work is provided at the end of this booklet. If you use this space you must clearly identify the question number you are attempting. Any rough work must be written in this booklet. You should score through your rough work when you have written your final copy.

Use **blue** or **black** ink.

Before leaving the examination room you must give this booklet to the Invigilator; if you do not, you may lose all the marks for this paper.

SECTION 1 — 20 marks

The questions for Section 1 are contained in the question paper X757/75/02.
Read these and record your answers on the answer grid on *Page three* opposite.
Use **blue** or **black** ink. Do NOT use gel pens or pencil.

1. The answer to each question is **either** A, B, C, D or E. Decide what your answer is, then fill in the appropriate bubble (see sample question below).
2. There is **only one correct** answer to each question.
3. Any rough work must be written in the additional space for answers and rough work at the end of this booklet.

Sample Question

The energy unit measured by the electricity meter in your home is the:

A ampere
B kilowatt-hour
C watt
D coulomb
E volt.

The correct answer is **B**—kilowatt-hour. The answer **B** bubble has been clearly filled in (see below).

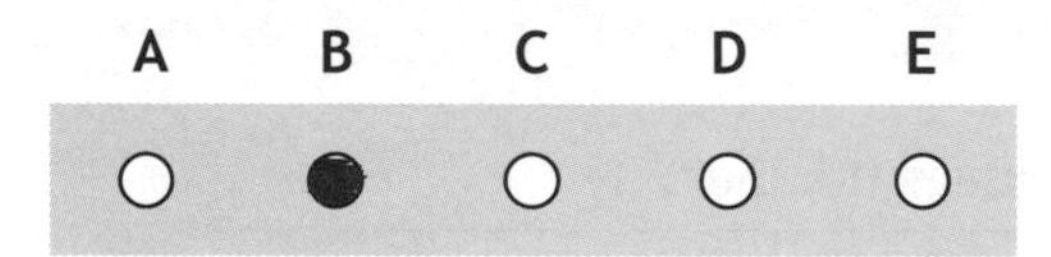

Changing an answer

If you decide to change your answer, cancel your first answer by putting a cross through it (see below) and fill in the answer you want. The answer below has been changed to **D**.

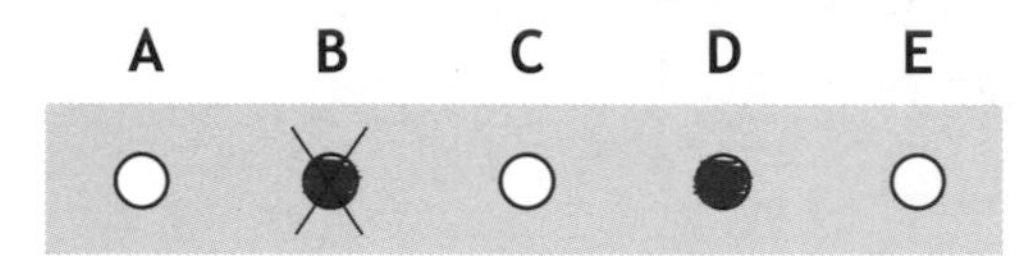

If you then decide to change back to an answer you have already scored out, put a tick (✓) to the **right** of the answer you want, as shown below:

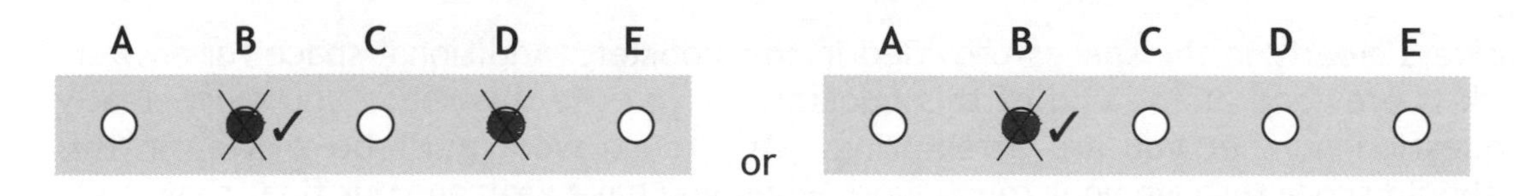

SECTION 1 — Answer Grid

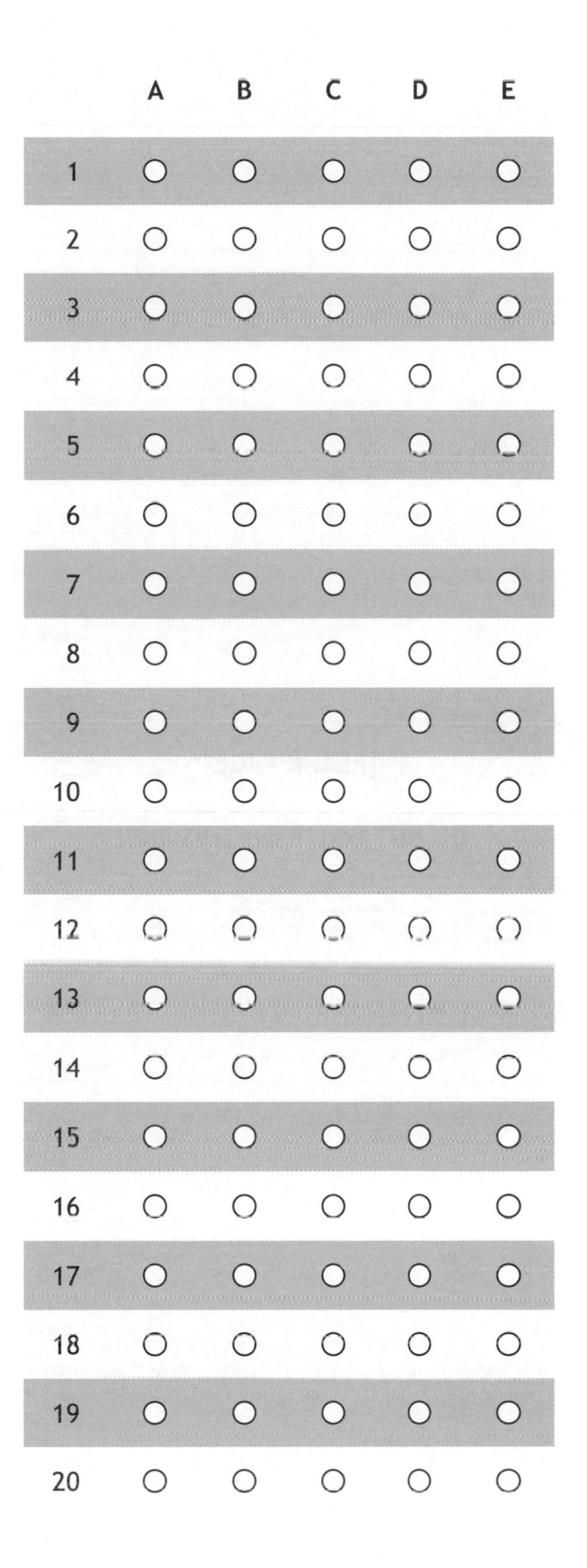

[BLANK PAGE]

DO NOT WRITE ON THIS PAGE

MARKS | DO NOT WRITE IN THIS MARGIN

[Turn over for Question 1 on *Page six*

DO NOT WRITE ON THIS PAGE

SECTION 2 — 90 marks

Attempt ALL questions

MARKS | DO NOT WRITE IN THIS MARGIN

1. A student sets up the following circuit using a battery, two lamps, a switch and a resistor.

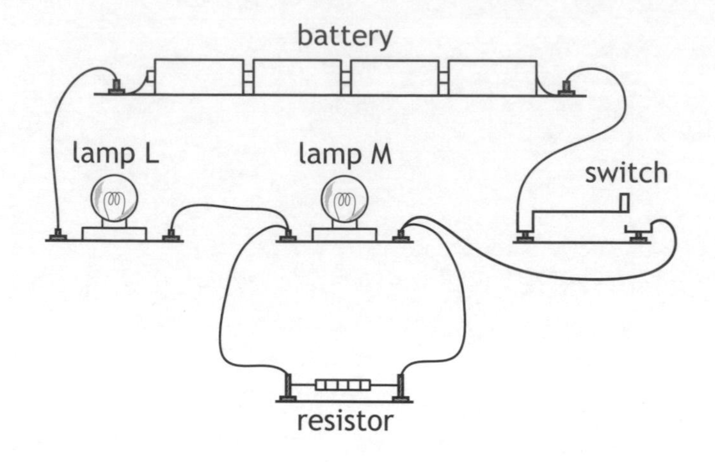

(a) Draw a circuit diagram for this circuit using the correct symbols for the components. **3**

(b) Each lamp is rated 2·5 V, 0·50 A.

Calculate the resistance of one of the lamps when it is operating at the correct voltage. **3**

Space for working and answer

MARKS DO NOT WRITE IN THIS MARGIN

1. **(continued)**

(c) When the switch is closed, will lamp L be brighter, dimmer or the same brightness as lamp M?

You **must** justify your answer. **3**

[Turn over

MARKS | DO NOT WRITE IN THIS MARGIN

2. (a) A student investigates the electrical properties of three different components; a lamp, an LED and a fixed resistor.

Current-voltage graphs produced from the student's results are shown.

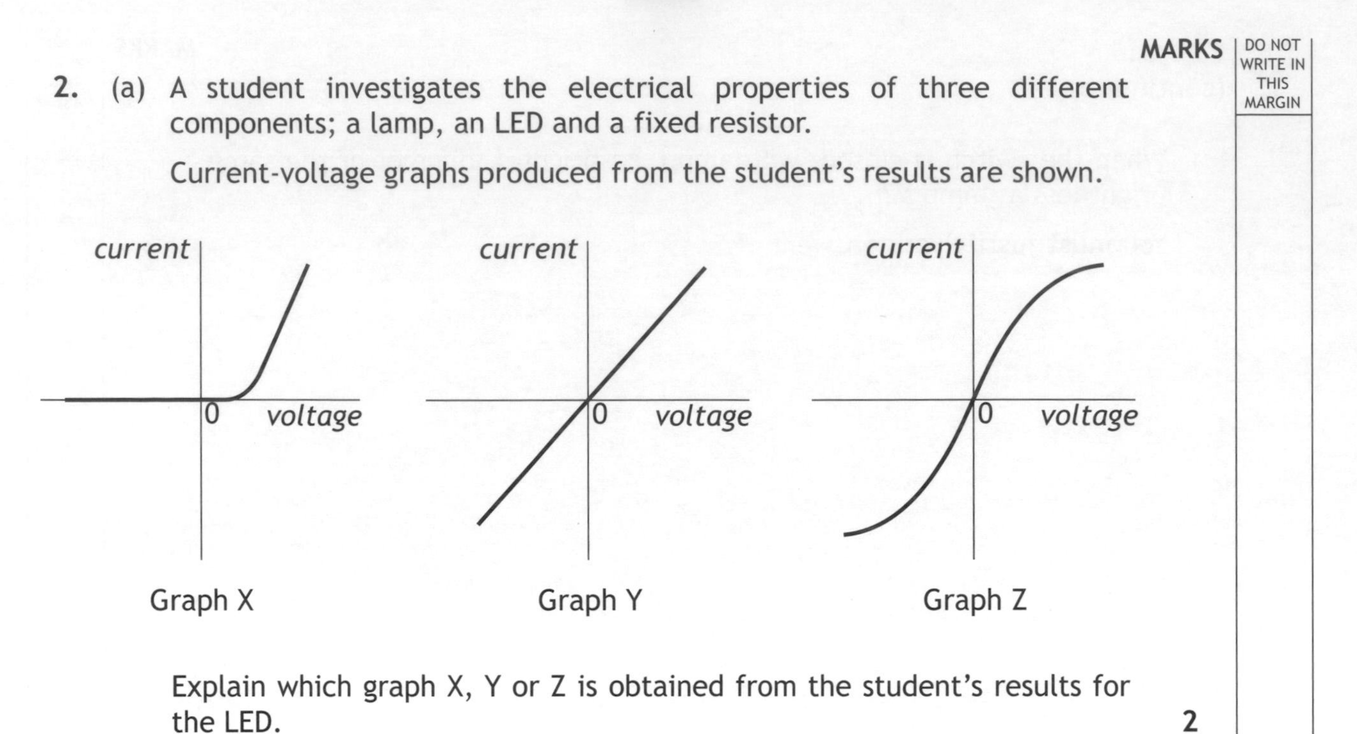

Explain which graph X, Y or Z is obtained from the student's results for the LED. **2**

(b) One of the components is operated at 4·0 V with a current of 0·50 A for 60 seconds.

(i) Calculate the energy transferred to the component during this time. **4**

Space for working and answer

MARKS DO NOT WRITE IN THIS MARGIN

2. (b) (continued)

(ii) Calculate the charge which passes through this component during this time. **3**

Space for working and answer

[Turn over

3. A technician uses pulses of ultrasound (high frequency sound) to detect imperfections in a sample of steel.

The pulses of ultrasound are transmitted into the steel.

The speed of ultrasound in steel is $5200\,m\,s^{-1}$.

Where there are no imperfections, the pulses of ultrasound travel through the steel and are reflected by the back wall of the steel.

Where there are imperfections in the steel, the pulses of ultrasound are reflected by these imperfections.

The reflected pulses return through the sample and are detected by the ultrasound receiver.

The technician transmits pulses of ultrasound into the steel at positions X, Y and Z as shown.

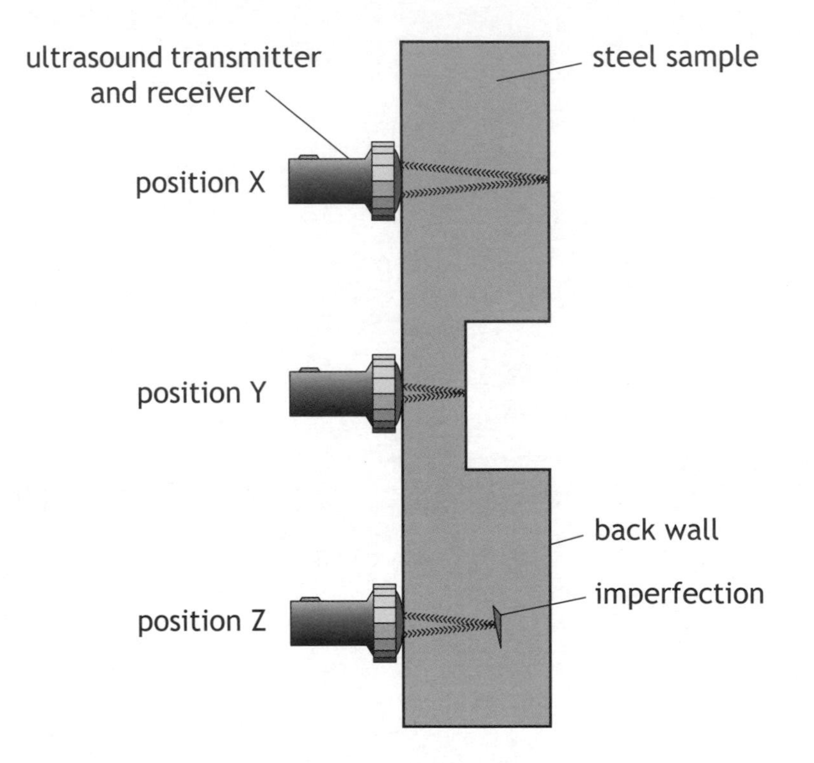

The times between the pulses being transmitted and received for positions X and Y are shown in the graph.

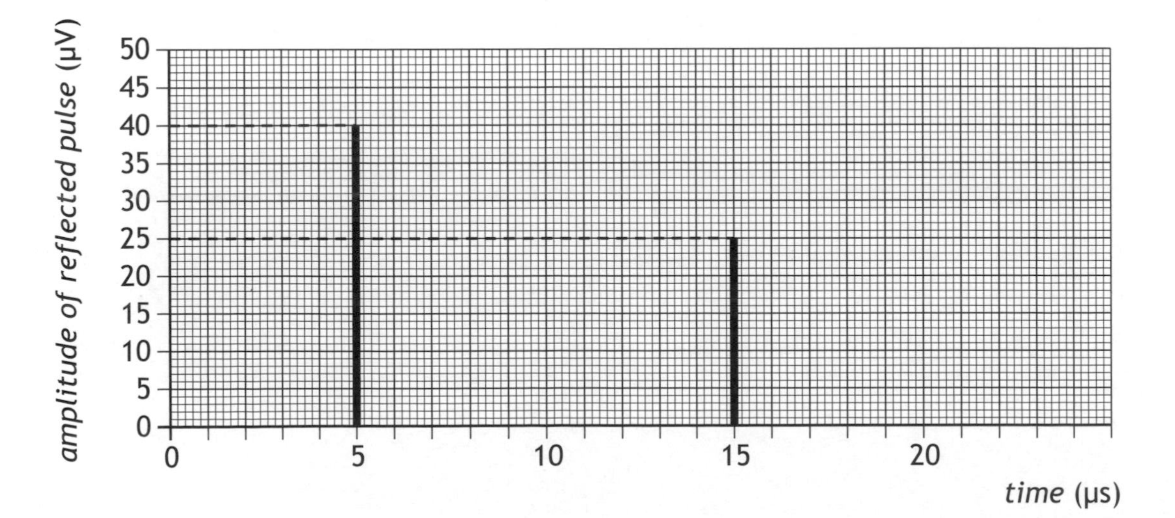

3. (continued)

MARKS DO NOT WRITE IN THIS MARGIN

(a) (i) State the time taken between the pulse being transmitted and received at position X. 1

(ii) Calculate the thickness of the steel sample at position X. 4

Space for working and answer

(b) On the graph on the previous page, draw a line to show the reflected pulse from position Z. 2

(c) The ultrasound pulses used have a period of 4·0 μs.

(i) Show that the frequency of the ultrasound pulses is $2{\cdot}5 \times 10^5$ Hz. 2

Space for working and answer

(ii) Calculate the wavelength of the ultrasound pulses in the steel sample. 3

Space for working and answer

MARKS DO NOT WRITE IN THIS MARGIN

3. (continued)

(d) The technician replaces the steel sample with a brass sample.

The brass sample has the same thickness as the steel sample at position X.

The technician transmits pulses of ultrasound into the brass at position P as shown.

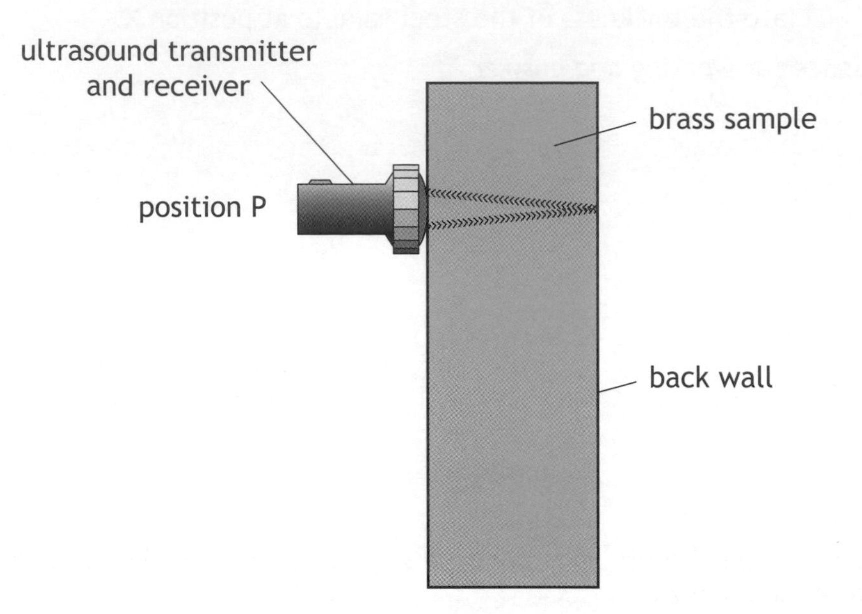

The time between the ultrasound pulse being transmitted and received at position P is greater than the time recorded at position X in the steel sample.

State whether the speed of ultrasound in brass is less than, equal to or greater than the speed of ultrasound in steel.

You **must** justify your answer. **2**

MARKS DO NOT WRITE IN THIS MARGIN

4. A science technician removes two metal blocks from an oven. Immediately after the blocks are removed from the oven the technician measures the temperature of each block, using an infrared thermometer. The temperature of each block is 230 °C.

After several minutes the temperature of each block is measured again. One block is now at a temperature of 123 °C and the other block is at a temperature of 187 °C.

Using your knowledge of physics, comment on possible explanations for this difference in temperature. **3**

[Turn over

MARKS DO NOT WRITE IN THIS MARGIN

5. Diamonds are popular and sought after gemstones.

Light is refracted as it enters and leaves a diamond.

The diagram shows a ray of light entering a diamond.

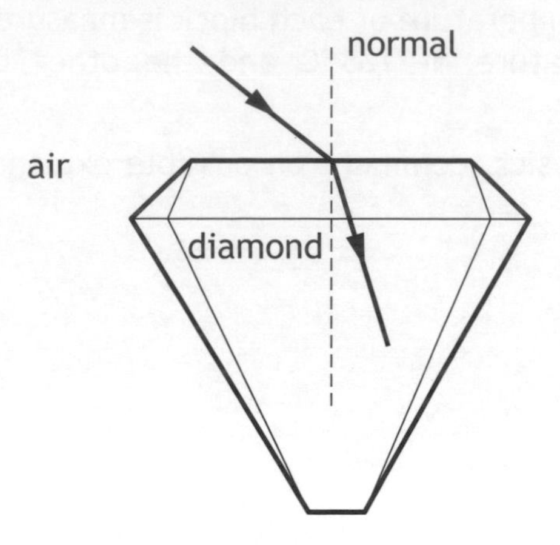

(a) On the diagram, label the angle of incidence *i* and the angle of refraction *r*. **1**

(b) State what happens to the speed of the light as it enters the diamond. **1**

(c) The optical density of a gemstone is a measure of its ability to refract light.

Gemstones of higher optical density cause more refraction.

A ray of light is directed into a gemstone at an angle of incidence of 45°.

The angle of refraction is then measured.

This is repeated for different gemstones.

Gemstone	*Angle of refraction*
A	24·3°
B	17·0°
C	27·3°
D	19·0°
E	25·5°

Diamond is known to have the highest optical density.

Identify which gemstone is most likely to be diamond. **1**

MARKS | DO NOT WRITE IN THIS MARGIN

5. (continued)

(d) Diamond is one of the hardest known substances.

Synthetic diamonds are attached to the cutting edges of drill bits for use in the oil industry.

These drill bits are able to cut into rock.

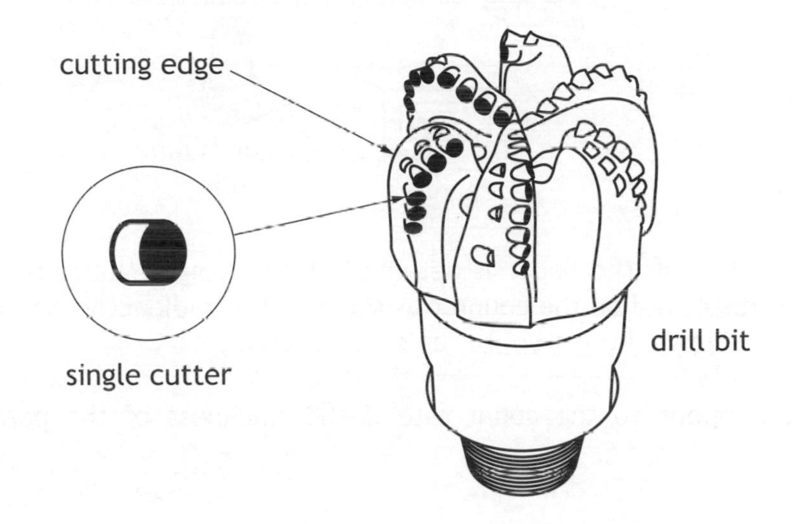

The area of a single cutter in contact with the rock is $1{\cdot}1 \times 10^{-5}\,m^2$.

When drilling, this cutter is designed to exert a maximum force of 61 kN on the rock.

Calculate the maximum pressure that the cutter can exert on the rock. **3**

Space for working and answer

[Turn over

MARKS DO NOT WRITE IN THIS MARGIN

6. A paper mill uses a radioactive source in a system to monitor the thickness of paper.

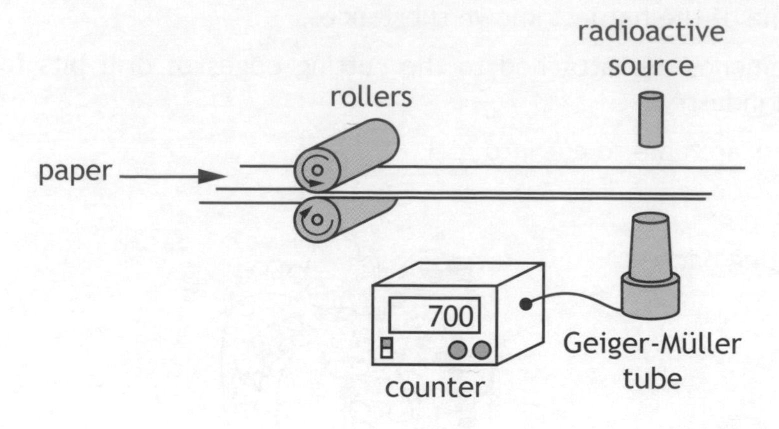

Radiation passing through the paper is detected by the Geiger-Müller tube. The count rate is displayed on the counter as shown. The radioactive source has a half-life that allows the system to run continuously.

(a) State what happens to the count rate if the thickness of the paper decreases. **1**

(b) The following radioactive sources are available.

Radioactive Source	*Half-life*	*Radiation emitted*
W	600 years	alpha
X	50 years	beta
Y	4 hours	beta
Z	350 years	gamma

(i) State which radioactive source should be used.

You **must** explain your answer. **3**

MARKS | DO NOT WRITE IN THIS MARGIN

6. (b) (continued)

(ii) State what is meant by the term *half-life*. 1

(iii) State what is meant by a gamma ray. 1

(c) The graph below shows how the activity of another radioactive source varies with time.

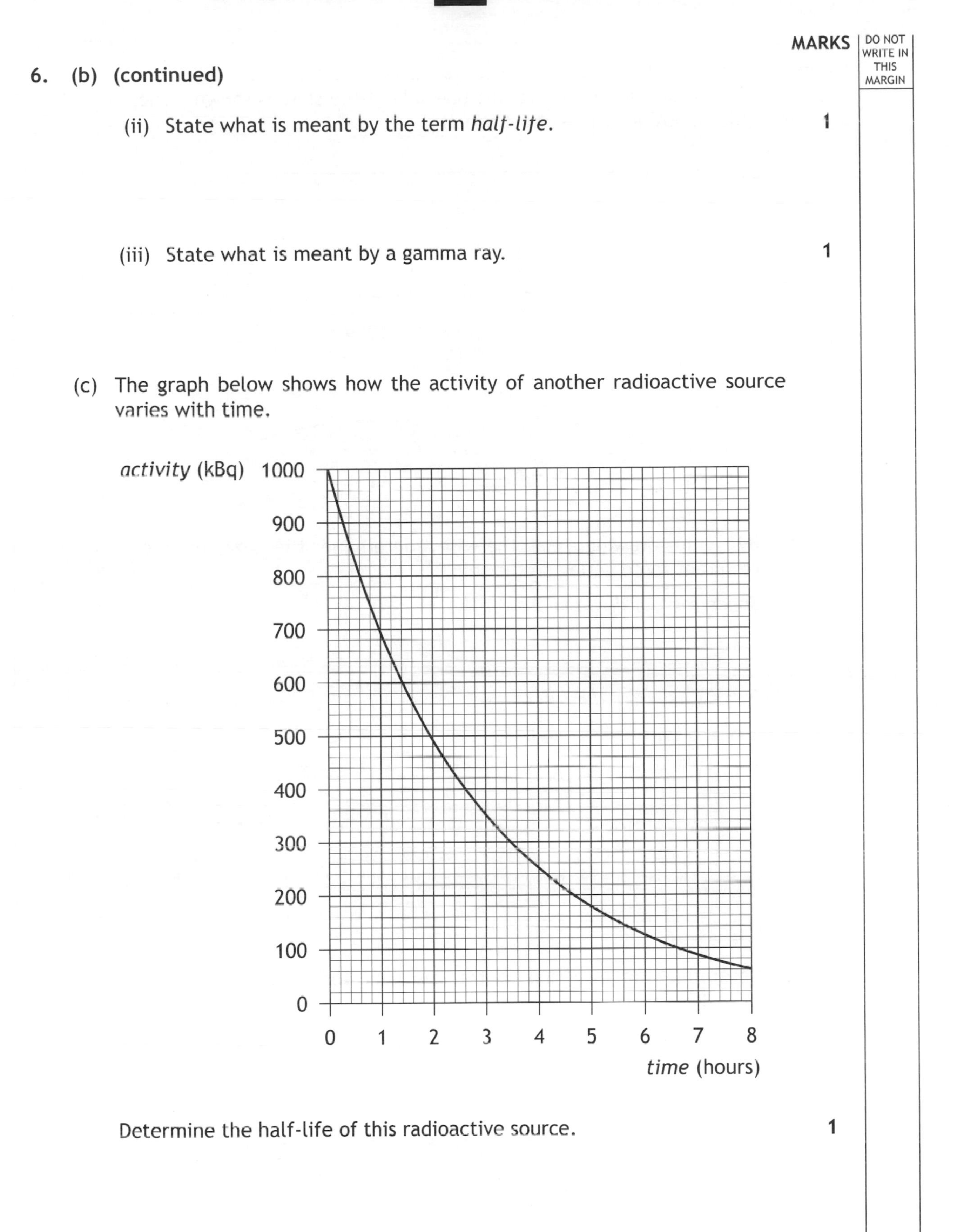

Determine the half-life of this radioactive source. 1

[Turn over

MARKS | DO NOT WRITE IN THIS MARGIN

7. A ship of mass $5{\cdot}0 \times 10^6$ kg leaves a port. Its engine produces a forward force of $8{\cdot}0 \times 10^3$ N. A tugboat pushes against one side of the ship as shown. The tugboat applies a pushing force of $6{\cdot}0 \times 10^3$ N.

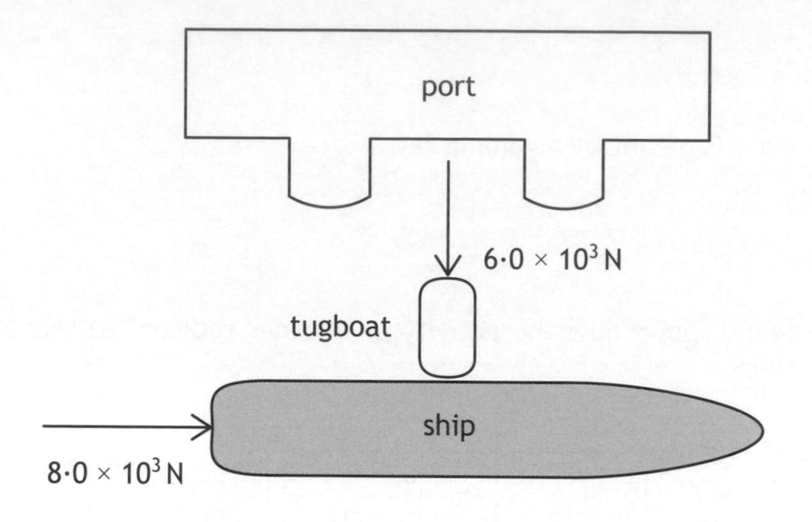

(a) (i) By scale drawing, or otherwise, determine the size of the resultant force acting on the ship. **2**

Space for working and answer

(ii) Determine the direction of the resultant force relative to the $8{\cdot}0 \times 10^3$ N force. **2**

Space for working and answer

MARKS DO NOT WRITE IN THIS MARGIN

7. (a) (continued)

(iii) Calculate the size of the acceleration of the ship. 3

Space for working and answer

(b) Out in the open sea the ship comes to rest.

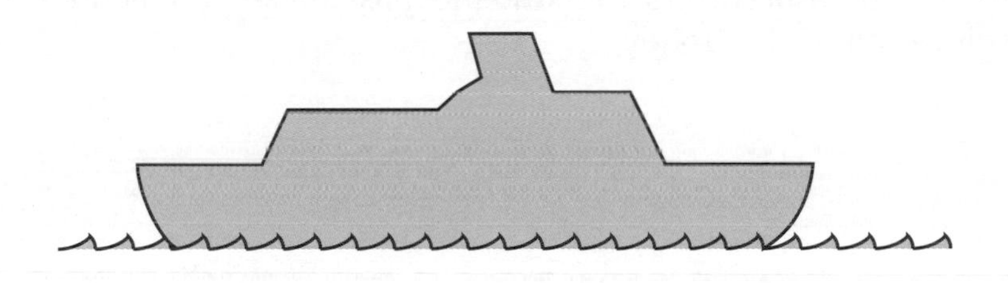

Explain, with the aid of a labelled diagram, why the ship floats. 3

[Turn over

MARKS DO NOT WRITE IN THIS MARGIN

8. A student is investigating the motion of a trolley down a ramp.

(a) The student uses the apparatus shown to carry out an experiment to determine the acceleration of a trolley as it rolls down a ramp.

The trolley is released from rest at the top of the ramp.

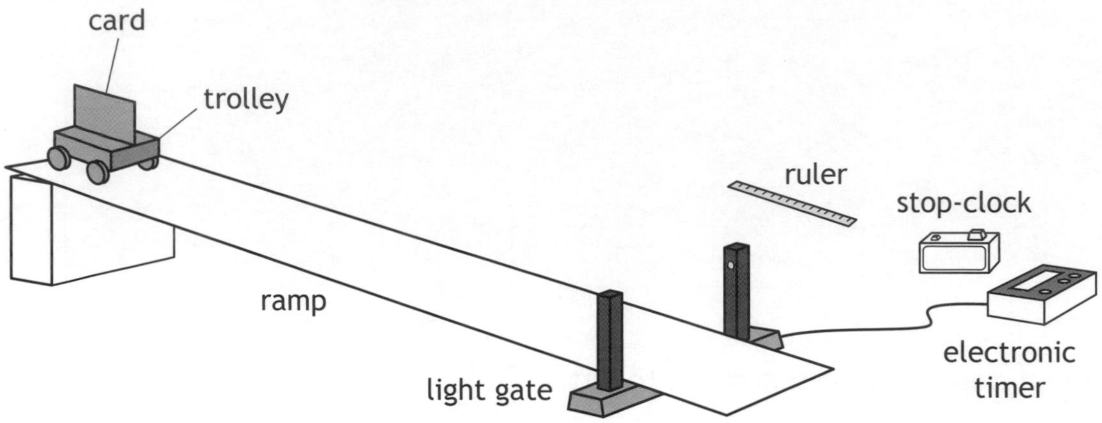

card
trolley
ruler
stop-clock
ramp
electronic timer
light gate

(i) State the measurements the student must make to calculate the acceleration of the trolley. **3**

(ii) Suggest one reason why the acceleration calculated from these measurements might not be accurate. **1**

MARKS DO NOT WRITE IN THIS MARGIN

8. (continued)

(b) In a second experiment, the student uses a motion sensor and computer to produce the following velocity-time graph for the trolley

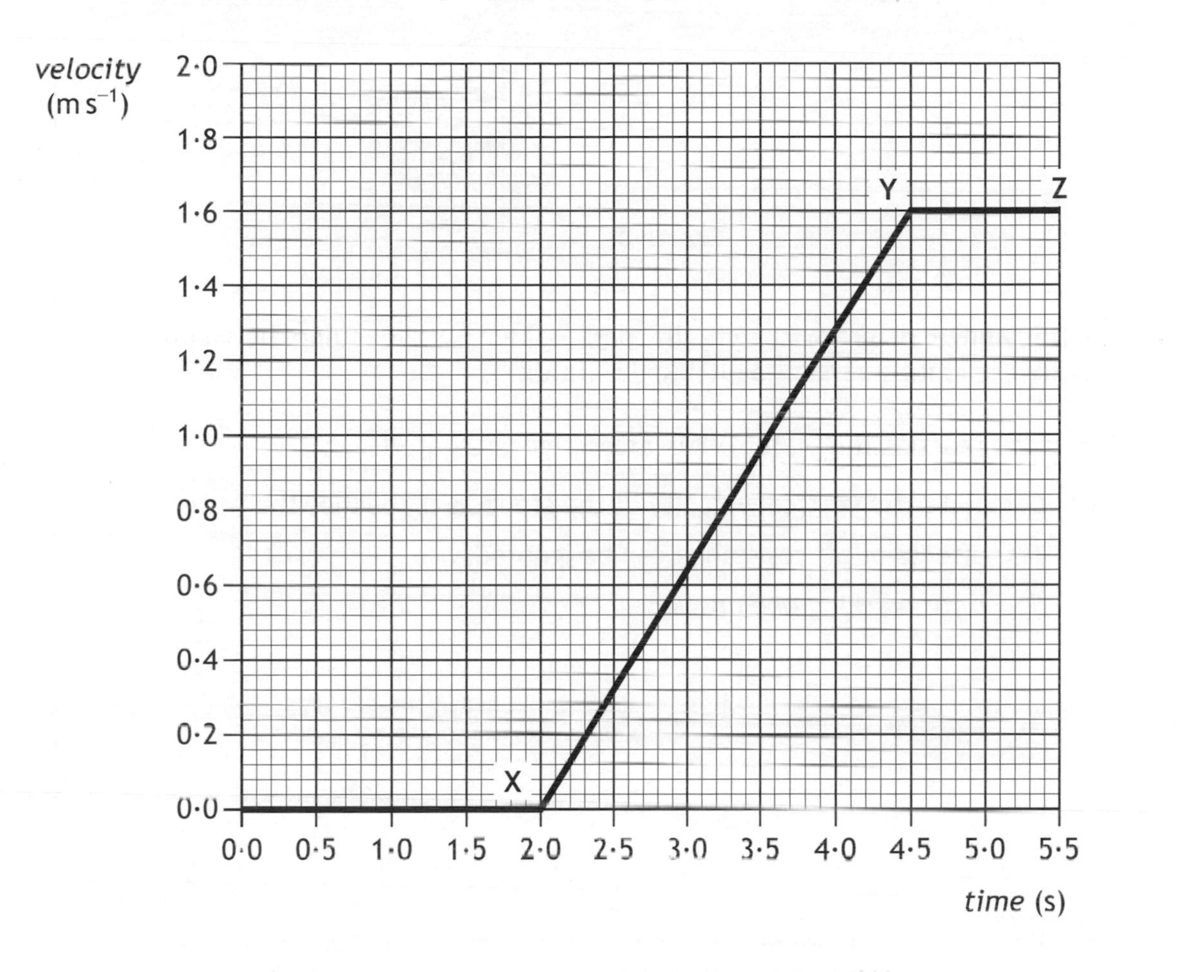

Calculate the acceleration of this trolley between X and Y. 3

Space for working and answer

[Turn over

MARKS | DO NOT WRITE IN THIS MARGIN

9. A child throws a stone horizontally from a bridge into a river.

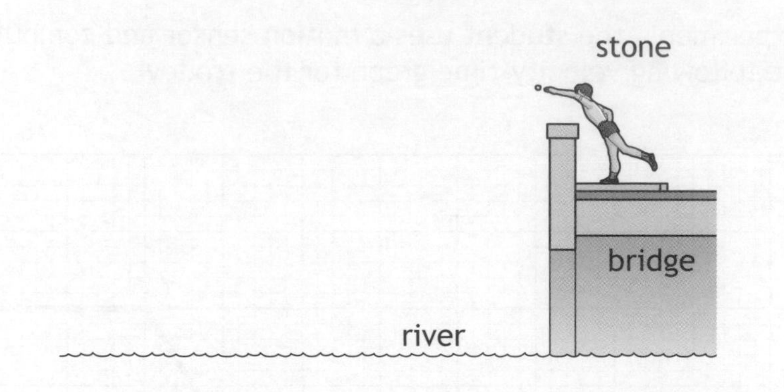

(a) On the above diagram sketch the path taken by the stone between leaving the child's hand and hitting the water. **1**

(b) The stone reaches the water 0·80 s after it was released.

(i) Calculate the vertical velocity of the stone as it reaches the water.

The effects of air resistance can be ignored. **3**

Space for working and answer

(ii) Determine the height above the water at which the stone was released. **4**

Space for working and answer

(c) The child now drops a similar stone vertically from the same height into the river.

State how the time taken for this stone to reach the water compares with the time taken for the stone in (b). **1**

MARKS DO NOT WRITE IN THIS MARGIN

10. Space exploration involves placing astronauts in difficult environments. Despite this, many people believe the benefits of space exploration outweigh the risks.

Using your knowledge of physics, comment on the benefits and/or risks of space exploration. **3**

[Turn over

[BLANK PAGE]

DO NOT WRITE ON THIS PAGE

MARKS DO NOT WRITE IN THIS MARGIN

11. Craters on the Moon are caused by meteors striking its surface.

A student investigates how a crater is formed by dropping a marble into a tray of sand.

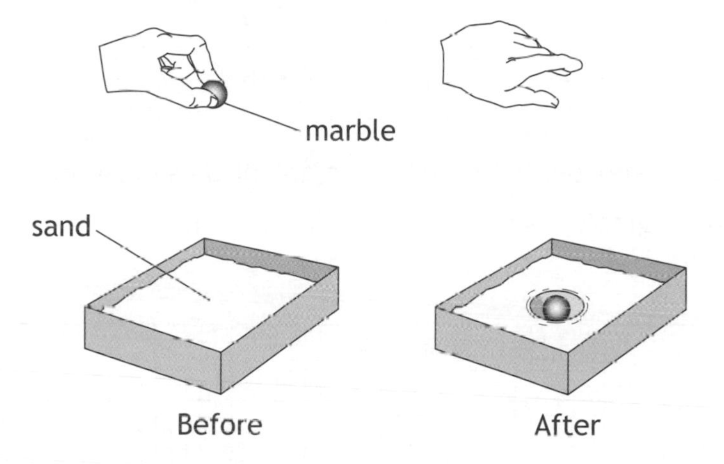

(a) The marble has a mass of 0·040 kg.

(i) Calculate the loss in potential energy of the marble when it is dropped from a height of 0·50 m. **3**

Space for working and answer

(ii) Describe the energy change that takes place as the marble hits the sand. **1**

[Turn over

MARKS DO NOT WRITE IN THIS MARGIN

11. (continued)

(b) The student drops the marble from different heights and measures the diameter of each crater that is formed.

The table shows the student's results.

height (m)	*diameter* (m)
0·05	0·030
0·10	0·044
0·15	0·053
0·35	0·074
0·40	0·076
0·45	0·076

(i) Using the graph paper below, draw a graph of these results. **3**

(Additional graph paper, if required, can be found on *Page twenty-eight*)

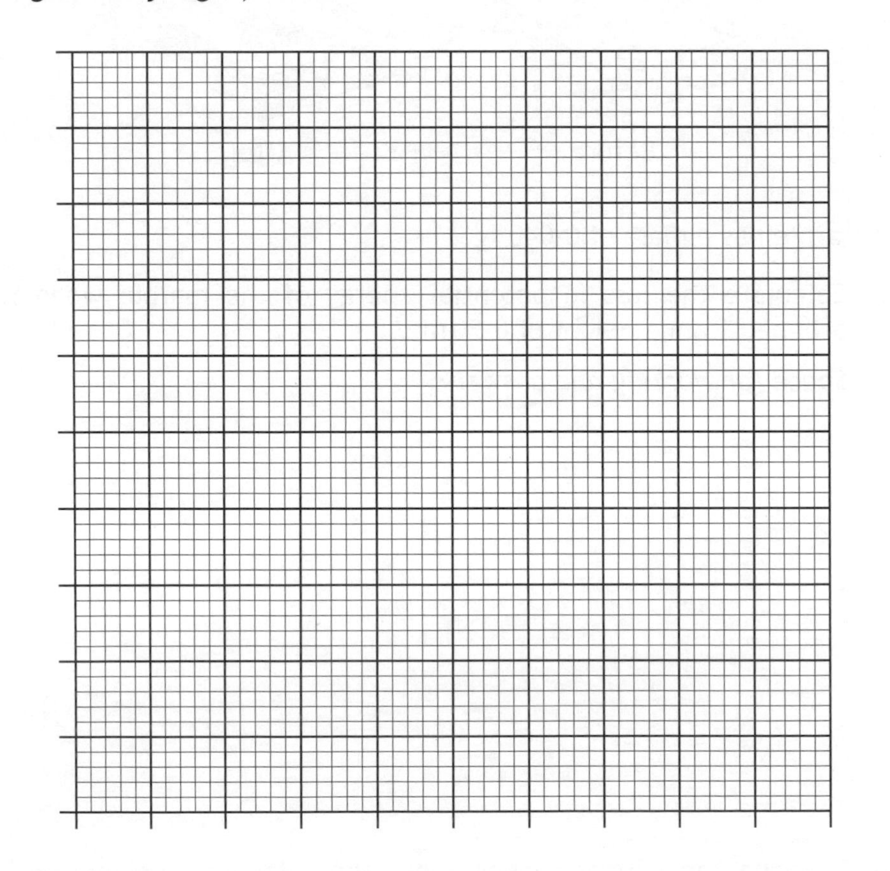

MARKS | DO NOT WRITE IN THIS MARGIN

11. (b) (continued)

(ii) Use your graph to predict the diameter of the crater that is formed when the marble is dropped from a height of 0·25 m. 1

(iii) Suggest two improvements that the student could make to this investigation. 2

(c) (i) Suggest another variable, which could be investigated, that may affect the diameter of a crater. 1

(ii) Describe experimental work that could be carried out to investigate how this variable affects the diameter of a crater. 2

[END OF QUESTION PAPER]

MARKS | DO NOT WRITE IN THIS MARGIN

ADDITIONAL SPACE FOR ANSWERS AND ROUGH WORKING

Additional graph paper for Q11 (b) (i)

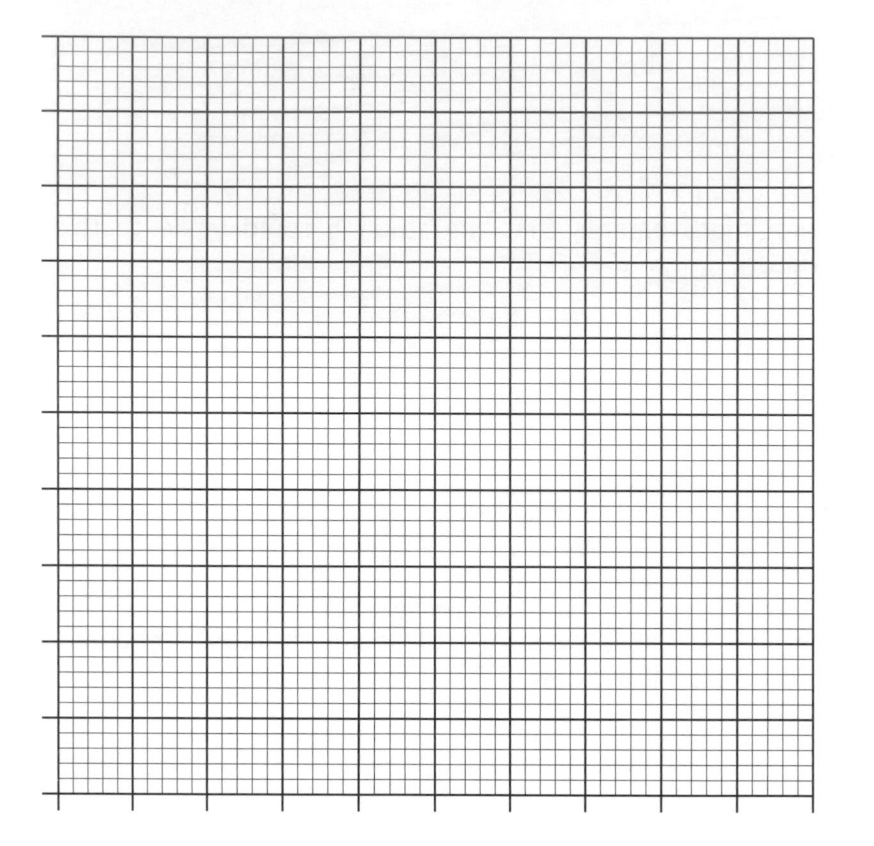

MARKS DO NOT WRITE IN THIS MARGIN

ADDITIONAL SPACE FOR ANSWERS AND ROUGH WORKING

MARKS DO NOT WRITE IN THIS MARGIN

ADDITIONAL SPACE FOR ANSWERS AND ROUGH WORKING

[BLANK PAGE]

DO NOT WRITE ON THIS PAGE

[BLANK PAGE]

DO NOT WRITE ON THIS PAGE

NATIONAL 5 | ANSWER SECTION

ANSWER SECTION FOR

SQA AND HODDER GIBSON NATIONAL 5 PHYSICS 2015

NATIONAL 5 PHYSICS MODEL PAPER 1

Section 1

1.	C	6.	C	11.	B	16.	B
2.	C	7.	A	12.	E	17.	D
3.	D	8.	B	13.	A	18.	D
4.	B	9.	E	14.	B	19.	A
5.	D	10.	A	15.	A	20.	C

Section 2

1. (a) $E_P = mgh$
$= 235 \times 9{\cdot}8 \times 12$
$= 27636$ J

(b) (i) $E_P = E_K$

$\frac{1}{2}mv^2 = mgh$

$\frac{1}{2} \times 2{\cdot}5 \times v^2 = 2{\cdot}5 \times 9{\cdot}8 \times 12$

$v = 15{\cdot}33 \text{ m s}^{-1}$

(ii) air resistance

2. (a) (i) I = 0·075 A
$V = IR$
$4{\cdot}2 = 0{\cdot}075 \cdot R$
$R = 56\ \Omega$

(ii) resistance stays the same because the graph is a straight line through the origin

(b) $\frac{1}{R_t} = \frac{1}{R_1} + \frac{1}{R_2}$

$= \frac{1}{33} + \frac{1}{56}$

$= 0{\cdot}048$

$R_t = 20{\cdot}76\ \Omega$

3. (a) Terminal velocity occurs when forces acting on a moving object become balanced.

(b) $F_d = 6\pi r\eta v_1$
$= 6 \times \pi \times 2{\cdot}83 \times 10^{-6} \times 1{\cdot}820 \times 10^{-5} \times 8{\cdot}56 \times 10^{-4}$
$= 8{\cdot}31 \times 10^{-13}\ N$

(c) $W = mg$
$8{\cdot}6 \times 10^{-13} = m \times 9{\cdot}8$
$m = 8{\cdot}8 \times 10^{-14}\ kg$

(d) a charged particle experiences a force in an electric field

4. (a) $E = cm\Delta T$
$= 902 \times 8000 \times (660 - 160)$
$= 3{\cdot}61 \times 10^9$ J

(b) $l_f = 3{\cdot}95 \times 10^5 \text{ J kg}^{-1}$
$E = ml$
$= 8000 \times 3{\cdot}95 \times 10^5$
$= 3{\cdot}16 \times 10^9$ J

5. Assume that the average mass of a student is 60 kg.

Student has a weight of W=mg=588 N.

The total weight of the student will be exerted on the ground where his feet are in contact.

Assume that the area of one foot in contact with ground is $0{\cdot}2 \text{ m} \times 0{\cdot}08 \text{ m} = 0{\cdot}016 \text{ m}^2$.

Total area in contact = $0{\cdot}032 \text{ m}^2$.

Pressure $= \frac{force}{area} = \frac{588}{0{\cdot}032} = 18\,375\ Pa$

6. (a) $P \times V$ = 2000, 1995, 2002, 2001

$P \times V$ = constant

(b) gas molecules collide with walls of container more often

so average force increases

causing an increase in pressure

(c) To reduce the inaccuracy of the syringe volume, since the volume of air contained in the tubing is not measured.

7. (a) Refraction occurs when light travels from one medium into another with a change in the wave speed.

(b) (i)

(ii)

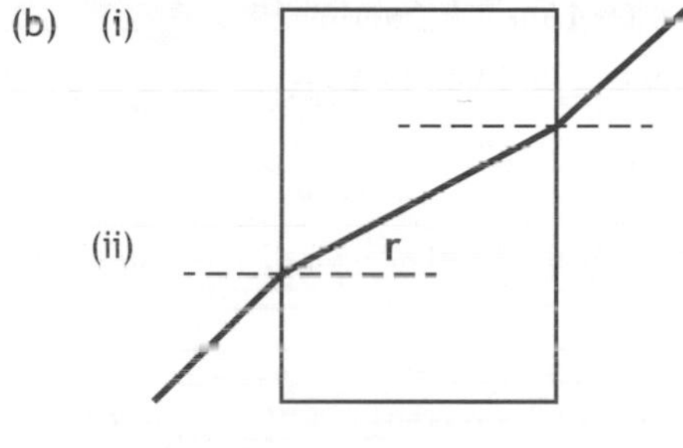

(c) (i) P ultraviolet

Q microwaves

(ii) $d = vt$

$4{\cdot}5 \times 10^{12} = 3 \times 10^8 \times t$

$t = 1{\cdot}5 \times 10^4$ s

8. (a) A particle containing two protons and two neutrons.

(b) Ionisation is the gain/loss of electrons by an atom.

(c) $4800 \xrightarrow{1} 2400 \xrightarrow{2} 1200 \xrightarrow{3} 600 \xrightarrow{4} 300$

4 half lives

$4 \cdot 2{\cdot}5 = 10$ hours

(d) Source may also emit β and/or γ radiation.

9. (a) (i) $D = \frac{E}{m}$

$= \frac{6 \times 10^{-6}}{0{\cdot}5}$

$= 1{\cdot}2 \times 10^{-5}\ Gy$

(ii) $H = Dw_R$

$= 1{\cdot}2 \times 10^{-5} \times 20$

$= 2{\cdot}4 \times 10^{-4}\ Sv$

(iii) $A = \frac{N}{t}$

$= \frac{24\,000}{(5 \times 60)}$

$= 80\ Bq$

(b) Fission.

10. (a) (i)

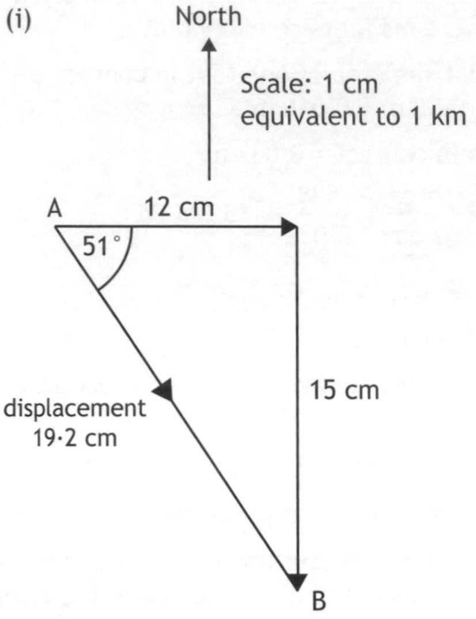

North

Scale: 1 cm equivalent to 1 km

A 12 cm

51°

15 cm

displacement 19·2 cm

B

Displacement is 19·2 km, bearing 141

(ii) $s = \bar{v}t$

$19{\cdot}2 = \bar{v} \times 1{\cdot}25$

$\bar{v} = 15{\cdot}4\ \text{km h}^{-1}$, bearing 141

(b) Displacement is 19·2 km, bearing 141

11. (a) $a = \frac{v - u}{t}$

$= \frac{3 - 0}{5}$

$= 0{\cdot}6\ \text{m s}^{-2}$

(b) $F = ma$

$= 40 \times 0{\cdot}6$

$= 24\ N$

12. (a) (i) 1·1 MW

(ii) power output not consistent

(b) different water speeds, different sizes of rotor blades

13. (a) Sagittarius A

(b) $d = vt$

$2{\cdot}6 \times 10^{20} = 3 \times 10^{8} \times t$

$t = 8{\cdot}7 \times 10^{11}$ seconds

$= \frac{8{\cdot}7 \times 10^{11}}{365{\cdot}25 \times 24 \times 60 \times 60}$

= 27 569 *light years*

(c) (i) Geiger-Müller tube

(ii) Compared to gamma rays, light rays have a **lower** frequency which means they have a **lower** energy.

14. If air resistance is ignored, then at the planet surface all objects fall with the **same** acceleration due to gravity g, so objects will take the same time to fall equal distances.

Different planets have different values for this acceleration g. The time taken to fall the same distance would be longer on planets where g is smaller.

On Earth, air resistance could reduce the acceleration of a falling object. If it was not streamlined, then an object would take longer to fall to the ground.

NATIONAL 5 PHYSICS MODEL PAPER 2

Section 1

1.	E	6.	E	11.	D	16.	D
2.	B	7.	C	12.	E	17.	E
3.	C	8.	B	13.	A	18.	E
4.	B	9.	C	14.	E	19.	C
5.	C	10.	E	15.	D	20.	B

Section 2

1. (a) $E_P = mgh$

$= 750 \times 9{\cdot}8 \times 7{\cdot}2$

$= 52\,920$ J

(b) (i) 52 920 J

(ii) $E_k = \frac{1}{2}mv^2$

$52920 = \frac{1}{2} \times 750 \times v^2$

$v = 11{\cdot}9$ m s^{-1}

2. (a) the resistance of LDR decreases when light level rises

voltage across R rises until MOSFET switches on the motor

(b) (i) 3000 Ω

(ii) $V_1 = \left(\frac{R_1}{R_1 + R_2}\right) V_s$

$V_1 = \left(\frac{600}{600 + 3000}\right) \times 12$

$V_1 = 2$ V

(iii) Since V is less than 2·4 V the transistor will not switch on so blinds do not shut.

3. (a) pressure is the force per unit area exerted on a surface

(b) $p = \rho gh$

$= 1025 \times 9{\cdot}8 \times 24$

$= 2{\cdot}4 \times 10^5$ *Pa*

4. Cars have to speed up regularly, reducing the overall mass of the car would reduce the unbalanced force required for acceleration (F=ma) reducing fuel needed.

Making the car more streamlined would reduce the frictional forces acting on the car, and so less fuel would be needed to provide the reduced forward force.

Design cars which transform their kinetic energy into electrical energy when braking, to be stored in rechargeable cells. These cells can provide an additional energy source to energise a motor which could assist the car's movement.

5. (a) $\frac{p}{T} = 347, 347, 346, 348, 348$

pressure and temperature are directly proportional when T is in Kelvin

(b) As temperature increases, the average E_k of the gas particles increases, so the particles collide with the walls of the container more frequently with greater force so pressure increases.

(c) To ensure all the gas in the flask is heated evenly.

6. (a) (i) $v = f\lambda$

$f = \frac{3 \times 10^8}{0{\cdot}06}$

$= 5 \times 10^9$ Hz

(ii) $T = \frac{1}{f}$

$= \frac{1}{5 \times 10^9}$

$= 2 \times 10^{-10}$ s

(b) signals received at same time
radio waves and microwaves have same speed

(c) The radio signals for car B have much higher frequency than for car A, so the radio signals for car B do not diffract as much around hills.

7. (a) Half-life is the time for the activity of a radioactive substance to reduce to half of its original value.

(b) *no. of half lives* $= \frac{\textit{no. of days}}{\textit{half life}} = \frac{13{\cdot}5}{2{\cdot}7} = 5$

64 [1]> 32 [2]> 16 [3]> 8 [4]> 4 [5]> 2

activity after 13·5 days = 2 *kBq*

(c) Minimise the time handling the sources, keep the radiation source at a maximum distance from the body.

8. (a) In a fission reaction, a nucleus with a large mass number splits into two nuclei of smaller mass numbers.

Energy is released and neutrons are usually released.

(b) $P = \frac{E}{t}$

$E = 1{\cdot}4 \times 10^9 \times 60 \times 60$

$E = 5{\cdot}0 \times 10^{12}$ (J)

Number of fissions $= \frac{5{\cdot}0 \times 10^{12}}{2{\cdot}9 \times 10^{-11}}$

$= 1{\cdot}7 \times 10^{23}$

(c) radioactive waste requires expensive storage

9. (a) $W = mg$

$= 50\,000 \times 9{\cdot}8$

$= 4{\cdot}9 \times 10^5$ N

(b) $4{\cdot}9 \times 10^5$ N

(c)

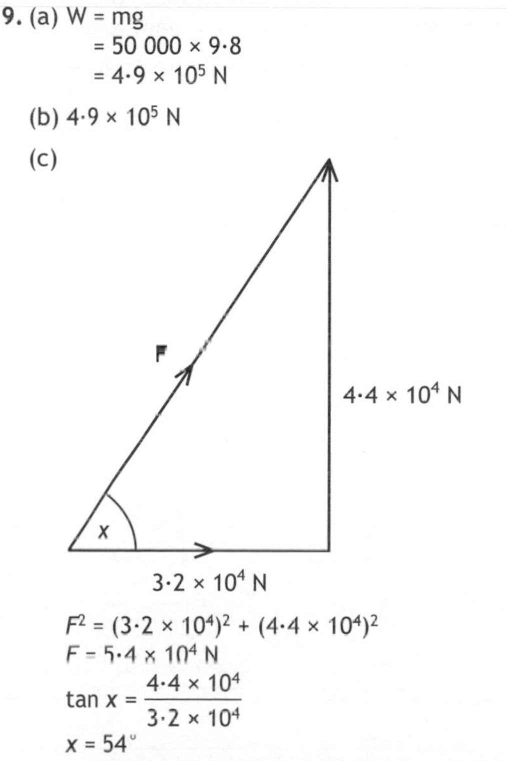

$F^2 = (3{\cdot}2 \times 10^4)^2 + (4{\cdot}4 \times 10^4)^2$

$F = 5{\cdot}4 \times 10^4$ N

$\tan x = \frac{4{\cdot}4 \times 10^4}{3{\cdot}2 \times 10^4}$

$x = 54°$

resultant force = $5{\cdot}4 \times 10^4$ N at bearing of 036

(d) $H = Dw_R$
$= 15 \times 10^{-6} \times 1$
$= 1{\cdot}5 \times 10^{-5}\ Sv$

(e) Ionisation is when an atom gains or loses electrons.

10. (a) $a = \frac{v - u}{t}$
$= \frac{9 - 0}{2}$
$a = 4{\cdot}5\ \text{m s}^{-2}$

(b) $F = ma$
$= 15 \times 4{\cdot}5$
$= 67{\cdot}5\ \text{N}$

(c) d = area under graph
$= (0{\cdot}5 \times 9 \times 2) + (10 \times 9) + (0{\cdot}5 \times 9 \times 1)$
$= 9 + 90 + 4{\cdot}5$
$= 103{\cdot}5\text{m}$

11. (a) $E_w = F \times d$
$= 300 \times 1{\cdot}5 \times 500$
$= 225\,000\ \text{J}$

(b) (i) $E = c\,m\,\Delta T$
$225\,000 = 902 \times 12 \times \Delta T$
$\Delta T = 21\ °\text{C}$

(ii) energy is lost to the surrounding air

12. As the rocket rises, fuel is used up and the rocket mass reduces. This reduces the rocket weight and so the unbalanced upward force increases, causing acceleration to increase as $a = \frac{F}{m}$.
As the rocket rises, the gravitational field strength decreases. This causes the rocket weight to reduce and so the unbalanced upward force increases, causing acceleration to increase as $a = \frac{F}{m}$.
As the rocket rises, the frictional force reduces because of the atmosphere being less dense at altitude, this reduces air resistance and so the unbalanced upward force increases, causing acceleration to increase as $a = \frac{F}{m}$.

13. (a) Neutron stars are thought to be formed when large stars collapse.

(b) During the collapsing process, electrons and protons combine to form neutrons.

(c) $T = \frac{1}{f}$
$= \frac{1}{716}$
$= 1{\cdot}4 \times 10^{-3}\ Hz$

NATIONAL 5 PHYSICS MODEL PAPER 3

Section 1

1.	B	**6.**	A	**11.**	C	**16.**	D
2.	D	**7.**	C	**12.**	E	**17.**	E
3.	D	**8.**	C	**13.**	A	**18.**	D
4.	C	**9.**	E	**14.**	D	**19.**	C
5.	E	**10.**	A	**15.**	B	**20.**	E

Section 2

1. (a) $E_p = mgh$
$= 0{\cdot}50 \times 9{\cdot}8 \times 19.3$
$= 94.6\ \text{J}$

(b) $Eh = cm\Delta T$
$94{\cdot}6 = 386 \times 0{\cdot}50 \times \Delta T$
$\Delta T = 0{\cdot}49\ °\text{C}$

(c) Temperature change is less.
Some heat energy is lost to the surroundings.

2. (a) $P = I^2 R$
$2 = I^2 \times 50$
$I^2 = 0{\cdot}04$
$I = 0{\cdot}2\ \text{A}$

(b) (i) $\frac{1}{R_T} = \frac{1}{R_1} + \frac{1}{R_2}$
$\frac{1}{R_T} = \frac{1}{60} + \frac{1}{30}$
$R_T = 20\ \Omega$

(ii) $P = \frac{V^2}{R}$
$P = \frac{9^2}{60}$
$= 1{\cdot}35\ \text{W}$
$P = \frac{V^2}{R}$
$P = \frac{9^2}{30}$
$= 2{\cdot}7\ \text{W}$

(iii) 30 Ω resistor will overheat

(c) none

3. The final temperature will be between both starting temperatures.

The heat energy E_h lost by the copper will be gained by the water.
From $\Delta T = \frac{E_h}{cm}$ the rise in temperature of the water and fall in temperature copper will depend on the respective masses of water and copper.

Specific heat capacity for water is much greater than for copper, so if masses were equal, final temperature would be closer to water's starting temperature.

4. (a) $p = \dfrac{F}{A}$

$F = 4{\cdot}6 \times 10^5 \times 3{\cdot}00 \times 10^{-2}$

$F = 13\,800$ N

(b) $P_1V_1 = P_2V_2$

$4{\cdot}6 \times 10^5 \times 1{\cdot}6 \times 10^{-3} = 1{\cdot}0 \times 10^5 \times V_2$

$V_2 = 7{\cdot}36 \times 10^{-3}$ m^3

(c) $V_{water} = V_2 - V_1 = 7{\cdot}36 \times 10^{-3} - 1{\cdot}6 \times 10^{-3} = 5{\cdot}76 \times 10^{-3}$ m^3

5. (a) $\lambda = \dfrac{distance}{no.\ of\ waves} = \dfrac{36}{8} = 4{\cdot}5$ cm $= 0{\cdot}045$ m

$v = f\lambda$

$= 5 \times 0{\cdot}045$

$= 0{\cdot}23$ m s^{-1}

(b)

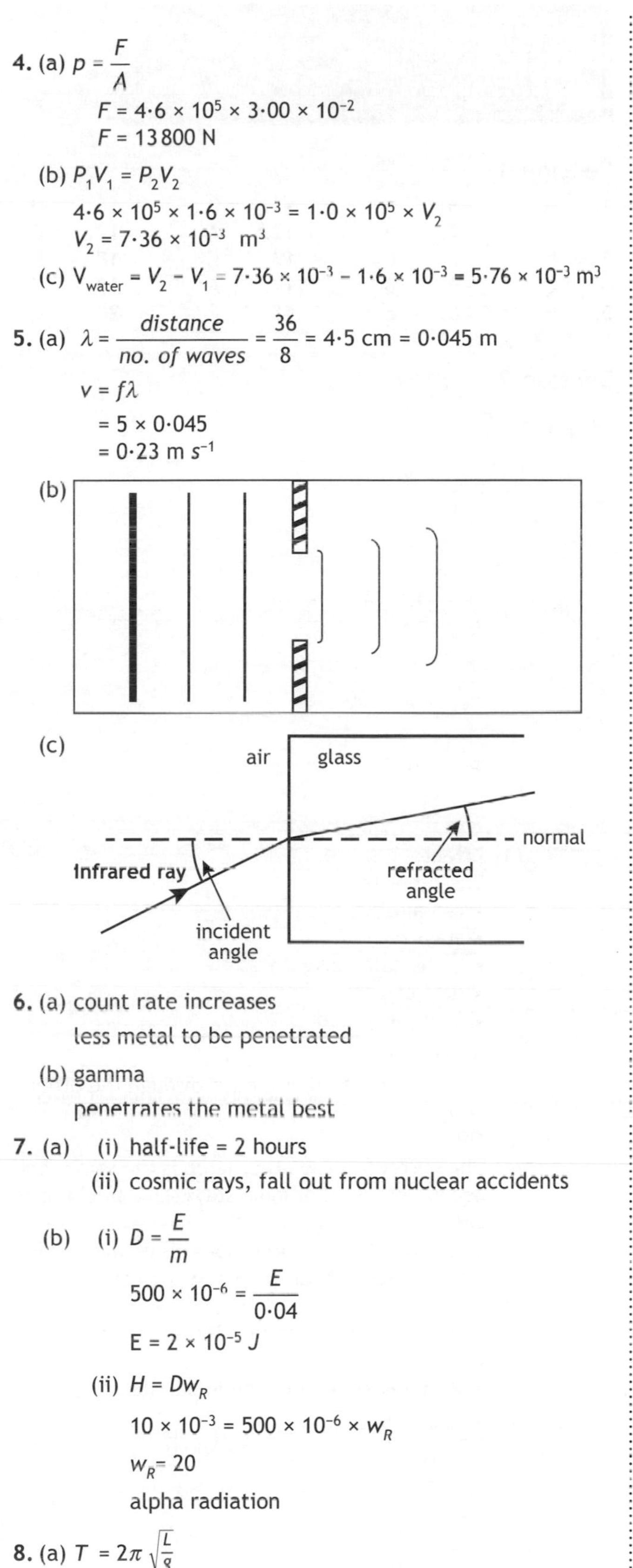

(c)

6. (a) count rate increases

less metal to be penetrated

(b) gamma

penetrates the metal best

7. (a) (i) half-life = 2 hours

(ii) cosmic rays, fall out from nuclear accidents

(b) (i) $D = \dfrac{E}{m}$

$500 \times 10^{-6} = \dfrac{E}{0{\cdot}04}$

$E = 2 \times 10^{-5}$ J

(ii) $H = Dw_R$

$10 \times 10^{-3} = 500 \times 10^{-6} \times w_R$

$w_R = 20$

alpha radiation

8. (a) $T = 2\pi\sqrt{\dfrac{L}{g}}$

$= 2\pi\sqrt{\dfrac{0{\cdot}8}{9{\cdot}8}}$

$= 1{\cdot}8$ s

(b) $T = \dfrac{1}{f}$

$1{\cdot}8 = \dfrac{1}{f}$

$f = 0{\cdot}6$ Hz

(c) no effect

9. (a)

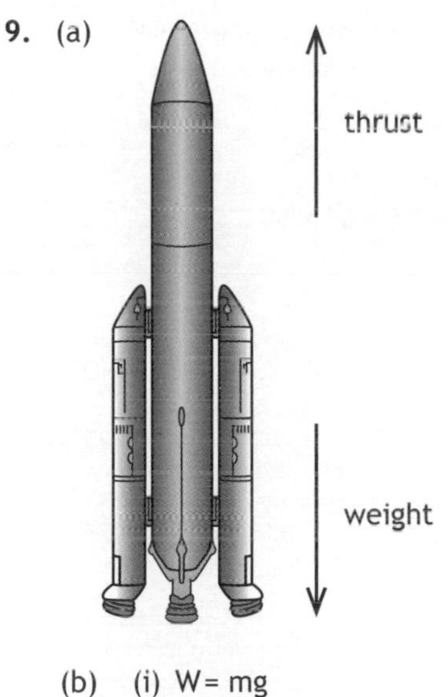

(b) (i) $W = mg$

$= 3{\cdot}08 \times 10^5 \times 9{\cdot}8$

$= 3{\cdot}02 \times 10^6$ N

(ii) $F_{unbalanced} = 3352 \times 10^3 - 3{\cdot}02 \times 10^6 = 332\,000$ N

$F = ma$

$332\,000 = 3{\cdot}08 \times 10^5 \times a$

$a = 1{\cdot}08$ m s^{-2}

(c) the ISS moves with constant speed in the horizontal direction

while accelerating due to the force of gravity in the vertical direction

(d) (i) The astronaut is falling towards Earth at the same rate as the ISS

(ii) The astronaut exerts a force against the wall

the wall exerts an equal and opposite force against the astronaut causing him to move.

10. (a) car continues at a <u>constant speed</u> during this time

AB represents driver's reaction time

(b) $E_k = \dfrac{1}{2}mv^2$

$= 0.5 \times 700 \times 30^2$

$= 315\,000$ J

(c) $a = \dfrac{v - u}{t}$

$= \dfrac{0 - 30}{2{\cdot}5}$

$= -12$ m s^{-2} = deceleration

$F = ma$

$= 700 \times 12$

$= 8400$ N

11. Racetrack has many bends and changes of direction which require both cars to decelerate and accelerate often.

Car B has highest acceleration which would allow it to reach higher speeds more quickly after a bend.

The higher maximum speed of car A is unlikely to be used because of the limited number of long straight parts of track before decelerating is required.

12. (a) Sunspots are caused by intense magnetic fields appearing beneath the sun's surface.

(b) ultraviolet radiation and X-rays

(c) $d = \bar{v}t$

$2{\cdot}28 \times 10^{11} = \bar{v} \times 2{\cdot}3 \times 24 \times 60 \times 60$

$\bar{v} = 1{\cdot}15 \times 10^{6}\ \text{m s}^{-1}$

NATIONAL 5 PHYSICS 2014

Section 1

1.	D	**6.**	A	**11.**	E	**16.**	D
2.	D	**7.**	A	**12.**	A	**17.**	D
3.	B	**8.**	C	**13.**	E	**18.**	E
4.	C	**9.**	B	**14.**	A	**19.**	C
5.	B	**10.**	B	**15.**	E	**20.**	D

Section 2

1. (a) $P = \dfrac{V^2}{R}$

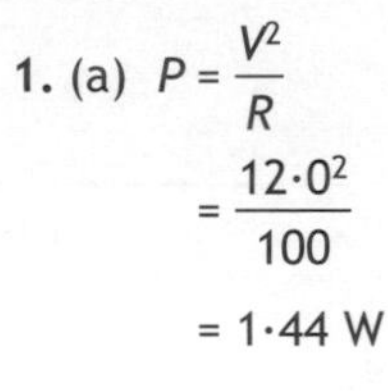

$= \dfrac{12{\cdot}0^2}{100}$

$= 1{\cdot}44\ \text{W}$

(b) (i) $\dfrac{1}{R_T} = \dfrac{1}{R_1} + \dfrac{1}{R_2} + \dfrac{1}{R_3}$

$\dfrac{1}{R_T} = \dfrac{1}{100} + \dfrac{1}{50} + \dfrac{1}{50}$

$\dfrac{1}{R_T} = \dfrac{1}{20}$

$R_T = 20\ \Omega$

(ii) **Effect:**

The other lamp:

- remains lit
- stays on
- is the same brightness
- gets brighter
- is not affected

Justification:

The current still has a path through the other lamp.

OR

The current in the other lamp is the same (only acceptable if other lamp stays same brightness).

OR

The current in the other lamp is greater (only acceptable if other lamp gets brighter).

OR

It has the same voltage/12 V (across it).

OR

The lamps are connected in parallel.

2. (a) (i) $V_2 = V_S - V_1 = 3{\cdot}0\ (\text{V})$

$I = \dfrac{V_2}{R}$

$= \dfrac{3{\cdot}0}{1050}$

$= (2{\cdot}857 \times 10^{-3}\ \text{A})$

$R_1 = \dfrac{V_1}{I}$

$= \dfrac{2{\cdot}0}{2{\cdot}857 \times 10^{-3}}$

$= 700\ \Omega$

(ii) 80 °C

(b) (i) As R_{th} increases, V_{th} increases
MOSFET/transistor turns on
Relay switches on (the heater).

(ii) Temperature decreases
Resistance of thermistor must be greater/increase to switch on MOSFET/transistor.

3. (a) $E = Pt$
$E = 15 \times 10 \times 60$
$E = 9000$ J

(b) (i) X

(ii) $E = cm\Delta T$
$9000 = c \times 1{\cdot}0 \times 10$
$c = 900$ J kg^{-1} °C^{-1}

(c) (i) Insulating the metal block
OR
Switch heater on for shorter time

(ii) Increase/greater (for insulating)
OR
Decrease/lower (for shorter time)

4. (a) f = *N° of waves/time*
$= \frac{4}{20}$
$= 0{\cdot}2$(Hz)
$v = f\lambda$
$= 0{\cdot}2 \times 12$
$= 2{\cdot}4$ m s^{-1}

(b)

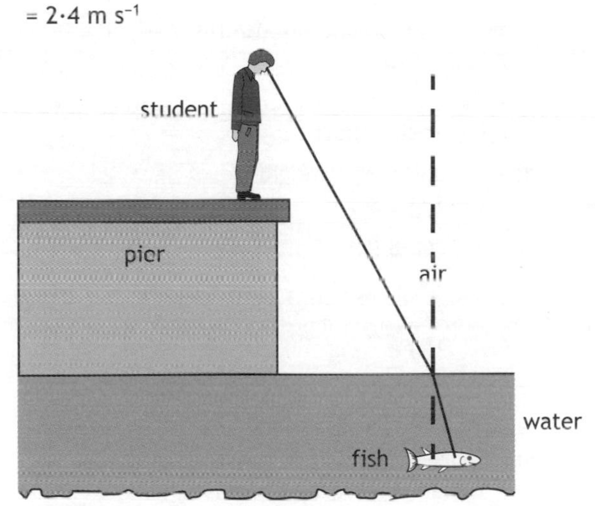

(1) mark for ray changing direction at water/air boundary.
(1) mark for angle in water less than angle in air. Angle of incidence in water should be less than the angle of refraction in air.
(1) mark for correct normal (must be placed at the point where a ray meets the water/air boundary).

5. (a) *UV index = (total effect of UV radiation x elevation above sea level adjustment x cloud adjustment) ÷ 25*

UV index = $(280 \times 1{\cdot}12 \times 0{\cdot}31) \div 25$
$= 3{\cdot}89$
$= 4$

(b)

	UVA	UVB	UVC
Type of sunscreen that absorbs most of this radiation	**P**	Q	**R**
Type of sunscreen that absorbs least of this radiation	R	**R**	**P**

(c) Detecting counterfeit bank notes, setting dental fillings, etc

6. (a) The time taken for the activity / corrected count rate of a radioactive source to half.

(b) (i) Measure the count in a set time interval
Repeat at regular intervals
Measure background count and subtract

(ii) Half-life = 10 minutes

(iii) 88 ⟶ 44 ⟶ 22 ⟶ 11 ⟶ 5·5
mark for evidence of halving
Count rate = 5·5 counts per minute

7. This is an open-ended question.
1 mark: The student has demonstrated a limited understanding of the physics involved. The student has made some statement(s) which is/are relevant to the situation, showing that at least a little of the physics within the problem is understood.
2 marks: The student has demonstrated a reasonable understanding of the physics involved. The student makes some statement(s) which is/are relevant to the situation, showing that the problem is understood.
3 marks: The maximum available mark would be awarded to a student who has demonstrated a good understanding of the physics involved. The student shows a good comprehension of the physics of the situation and has provided a logically correct answer to the question posed. This type of response might include a statement of the principles involved, a relationship or an equation, and the application of these to respond to the problem. This does not mean the answer has to be what might be termed an "excellent" answer or a "complete" one.

8. (a) (i) $D = \frac{E}{m}$
$= \frac{7{\cdot}2 \times 10^{-3}}{80{\cdot}0}$
$= 9{\cdot}0 \times 10^{-5}$ Gy

(ii) $H = Dw_R$
$= 9{\cdot}0 \times 10^{-5} \times 1$
$= 9{\cdot}0 \times 10^{-5}$ Sv

(b) When an **atom** gains or loses electrons.

9. This is an open-ended question.

1 mark: The student has demonstrated a limited understanding of the physics involved. The student has made some statement(s) which is/are relevant to the situation, showing that at least a little of the physics within the problem is understood.

2 marks: The student has demonstrated a reasonable understanding of the physics involved. The student makes some statement(s) which is/are relevant to the situation, showing that the problem is understood.

3 marks: The maximum available mark would be awarded to a student who has demonstrated a good understanding of the physics involved. The student shows a good comprehension of the physics of the situation and has provided a logically correct answer to the question posed. This type of response might include a statement of the principles involved, a relationship or an equation, and the application of these to respond to the problem. This does not mean the answer has to be what might be termed an "excellent" answer or a "complete" one.

10. (a) (i) $a = \frac{v - u}{t}$

$= \frac{4{\cdot}8 - 0}{25}$

$= 0{\cdot}19 \text{ m s}^{-2}$

(ii) constant speed

OR

constant velocity

(iii)

friction ← | boat | → forward force

OR

forward force ← | boat | → friction

(b) (i) distance = area under graph

$= \left(\frac{1}{2} \times 25 \times 4{\cdot}8\right) + (4{\cdot}8 \times 425) + \left(\frac{1}{2} \times 60 \times 4{\cdot}8\right)$

(= 60 + 2040 + 144)

= 2244 m

(ii) $v = \textit{total distance/time}$

= 2244/510

$= 4{\cdot}4 \text{ m s}^{-1}$

11. (a) To check that the maximum take-off weight is not exceeded.

(b) 19 625 N

(c) $d = vt$

$201\,000 = 67 \times t$

$t = 3000 \text{ s}$

12. (a) $W = mg$

$= 0{\cdot}94 \times 9{\cdot}8$

$= 9{\cdot}2 \text{ N}$

(b) **Method 1**

$A = 3 \times (2{\cdot}0 \times 10^{-4})$

$= 6{\cdot}0 \times 10^{-4} \text{ (m}^2\text{)}$

$p = \frac{F}{A}$

$= \frac{9{\cdot}2}{6{\cdot}0 \times 10^{-4}}$

$= 1{\cdot}5 \times 10^4 \text{ Pa}$

OR

Method 2

$p = \frac{F}{A}$

$= \frac{9{\cdot}2}{2{\cdot}0 \times 10^{-4}}$

$= 4{\cdot}6 \times 10^4 \text{ Pa}$

(If this line is the candidate's final answer, unit required.)

total $p = \frac{4{\cdot}6 \times 10^4}{A}$

$= 1{\cdot}5 \times 10^4 \text{ Pa}$

OR

Method 3...

take ⅓ of weight and use this for F in $p = F/A$

(c) Rocket/bottle pushes down on water, water pushes up on rocket/bottle.

(d) F_{un} = upthrust − weight

$= 370 - 9{\cdot}2$

$= 360{\cdot}8 \text{ (N)}$

$a = \frac{F}{m}$

$= \frac{360{\cdot}8}{0{\cdot}94}$

$= 380 \text{ m s}^{-2}$

(e)
- more water will increase weight/mass
- unbalanced force decreases
- acceleration is less

NATIONAL 5 PHYSICS 2015

Section 1

1.	A	6.	D	11.	E	16.	C
2.	A	7.	D	12.	A	17.	A
3.	C	8.	A	13.	E	18.	B
4.	E	9.	C	14.	C	19.	E
5.	B	10.	E	15.	B	20.	D

Section 2

1. (a) 2 marks for symbols:

- All correct 2
- At least two different symbols correct 1

1 mark for correct representation of external circuit wiring with no gaps

(b) $V = IR$ 1

$2{\cdot}5 = 0{\cdot}5 \times R$ 1

$R = 5\,\Omega$

(c) Mark for effect can only be awarded if a justification is attempted.

Incorrect or no effect stated, regardless of justification – no marks.

Effect:

(It/lamp L is) brighter 1

Justification:

M is in parallel (with resistor) 1

Greater current in/through lamp L (than that in M) 1

OR

Effect:

(It/lamp L is) brighter 1

Justification:

M is in parallel (with resistor) 1

Greater voltage across lamp L (than across M) 1

2. (a) (Graph) X 1

An LED/diode/it only conducts in one direction 1

(b) (i) $P = IV$ 1

$P = 0{\cdot}5 \times 4$

$P = 2$(W)

$E = Pt$ 1

$E = 2 \times 60$ 1

$E = 120$ J 1

(ii) $Q = I \times t$ 1

$Q = 0{\cdot}5 \times 60$ 1

$Q = 30$ C 1

3. (a) (i) 15 μs

(ii) **Method 1:**

$d = vt$ 1

$= 5200 \times 15 \times 10^{-6}$ 1

$= 0{\cdot}078$ (m) 1

(If this line is the final answer then unit required for mark)

$thickness = \frac{0{\cdot}078}{2}$

$= 0{\cdot}039$ m 1

Method 2:

$time = \frac{15 \times 10^{-6}}{2}$

$= 7{\cdot}5 \times 10^{-6}$ (s) 1

$d = vt$ 1

$= 5200 \times 7{\cdot}5 \times 10^{-6}$ 1

$= 0{\cdot}039$ m 1

(b)

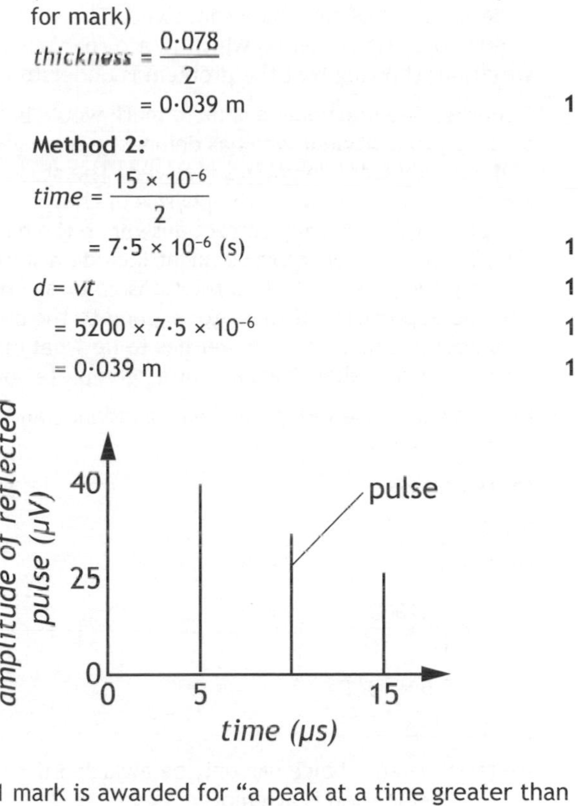

1 mark is awarded for "a peak at a time greater than 5 μs and less than 15 μs"

and

1 mark is awarded for "an amplitude greater than 25 μV and less than 40 μV"

(c) (i) This is a "show that" question so must start with correct formula or zero marks.

$f = \frac{1}{T}$ 1

$= \frac{1}{4{\cdot}0 \times 10^{-6}}$ 1

$= 2{\cdot}5 \times 10^{5}$ Hz

(ii) $v = f\lambda$ 1

$5200 = 2{\cdot}5 \times 10^{5} \times \lambda$ 1

$\lambda = 0{\cdot}021$ m 1

(d) Mark for effect can only be awarded if a justification is attempted.

Incorrect or no effect stated, regardless of justification – no marks.

(Speed of ultrasound in brass is) less (than in steel). 1

Takes greater time to travel (same) distance/thickness. 1

4. Demonstrates no understanding 0 marks

Demonstrates limited understanding 1 mark

Demonstrates reasonable understanding 2 marks

Demonstrates good understanding 3 marks

This is an open-ended question.

1 mark: The student has demonstrated a limited understanding of the physics involved. The student has made some statement(s) which is/are relevant to the situation, showing that at least a little of the physics within the problem is understood.

2 marks: The student has demonstrated a reasonable understanding of the physics involved. The student makes some statement(s) which is/are relevant to the situation, showing that the problem is understood.

3 marks: The maximum available mark would be awarded to a student who has demonstrated a good understanding of the physics involved. The student shows a good comprehension of the physics of the situation and has provided a logically correct answer to the question posed. This type of response might include a statement of the principles involved, a relationship or an equation, and the application of these to respond to the problem. This does not mean the answer has to be what might be termed an "excellent" answer or a "complete" one.

5. (a) Correctly labelled the angle of incidence **and** angle of refraction

(b) Decreases

(c) B

(d) $P = \frac{F}{A}$ 1

$= \frac{61000}{1 \cdot 1 \times 10^{-5}}$ 1

$= 5 \cdot 5 \times 10^{9}$ Pa 1

6. (a) Increases

(b) (i) Mark for choice can only be awarded if an explanation is attempted.

Incorrect or no choice made, regardless of explanation — no marks.

Choice:

(source) X 1

Explanation:

beta (source required) 1

long half-life 1

(ii) Time for activity to (decrease by) half

OR

Time for half the nuclei to decay

(iii) (high frequency) electromagnetic wave

(c) 2 hours

7. (a) (i) **Using Pythagoras:**

$\text{Resultant}^2 = (6 \cdot 0 \times 10^3)^2 + (8 \cdot 0 \times 10^3)^2$ 1

Resultant = 10×10^3 N 1

Using scale diagram:

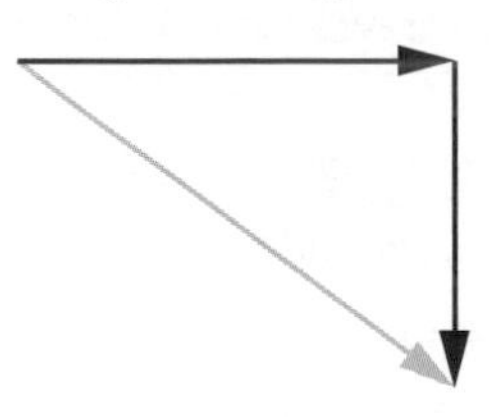

vectors to scale 1

Resultant = 10×10^3 N 1

(allow $\pm 0 \cdot 5 \times 10^3$ N tolerance)

(ii) **Using trigonometry:**

$\tan \theta = 6/8$ 1

$\theta = 37°$ 1

Using scale diagram:

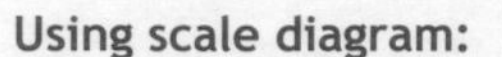

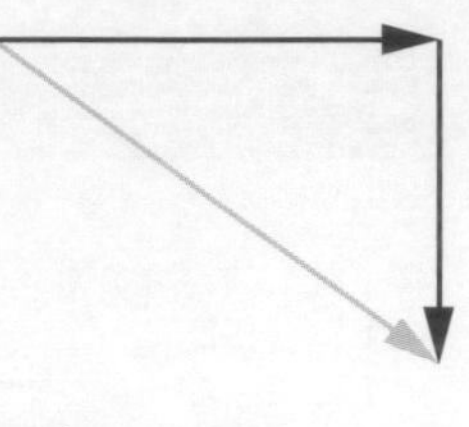

angles correct 1

$\theta = 37°$ 1

(allow $\pm 2°$ tolerance)

(iii) $F = ma$ 1

$10 \times 10^3 = 5 \cdot 0 \times 10^6 \times a$ 1

$a = 2 \cdot 0 \times 10^{-3}$ ms^{-2} 1

(b) **Upward arrow:** buoyancy force/upthrust/force of water on ship/flotation force 1

Downward arrow: weight/force of gravity 1

(These) forces are balanced 1

8. (a) (i) • length/width of card 1

• time taken for card to pass (through) the light gate 1

• time taken (for trolley to travel from starting position) to light gate 1

(ii) reaction time (can cause error with the stop clock reading)

OR

card may not have passed straight through light gate

OR

length/width of card not measured properly (e.g. ruler not straight along card)

OR

other suitable reason

(b) $a = \frac{v - u}{t}$ 1

$= \frac{1 \cdot 6 - 0}{2 \cdot 5}$ 1

$= 0 \cdot 64$ ms^{-2} 1

9. (a) suitable curved path 1

(b) (i) $a = \frac{v - u}{t}$ 1

$9 \cdot 8 = \frac{v - 0}{0 \cdot 80}$ 1

$v = 7 \cdot 8$ ms^{-1} 1

(ii) $\bar{v} = 3 \cdot 9$ ms^{-1} 1

$d = \bar{v}t$ 1

$= 3 \cdot 9 \times 0 \cdot 80$ 1

$= 3 \cdot 1$ m 1

(c) (it will take the) same (time)

10. Demonstrates no understanding 0 marks

Demonstrates limited understanding 1 mark

Demonstrates reasonable understanding 2 marks

Demonstrates good understanding 3 marks

This is an open-ended question.

1 mark: The student has demonstrated a limited understanding of the physics involved. The student has made some statement(s) which is/are relevant to the situation, showing that at least a little of the physics within the problem is understood.

2 marks: The student has demonstrated a reasonable understanding of the physics involved. The student makes some statement(s) which is/are relevant to the situation, showing that the problem is understood.

3 marks: The maximum available mark would be awarded to a student who has demonstrated a good understanding of the physics involved. The student shows a good comprehension of the physics of the situation and has provided a logically correct answer to the question posed. This type of response might include a statement of the principles involved, a relationship or an equation, and the application of these to respond to the problem. This does not mean the answer has to be what might be termed an "excellent" answer or a "complete" one.

11. (a) (i) $E_p = mgh$ 1

$E_p = 0{\cdot}040 \times 9{\cdot}8 \times 0{\cdot}50$ 1

$E_p = 0{\cdot}20$ J 1

(ii) kinetic (energy) to heat (and sound)

OR

kinetic (energy) of the marble to kinetic (energy) of the sand.

(b) (i) suitable scales, labels and units 1

all points plotted accurately to ± half a division 1

best fit <u>curve</u> 1

(ii) Consistent with best fit curve from (b)(i).

(iii) Any two from:

- Repeat (and average)
- Take (more) readings in the 0·15 (m) to 0·35 (m) drop height range
- Increase the height range
- level sand between drops
- or other suitable improvement

(1) each

(c) (i) suitable variable

e.g.

- mass/weight of marble
- angle of impact
- type of sand
- diameter of marble
- radius of marble
- density of marble
- volume of marble
- speed of marble
- time of drop

(ii) How independent variable can be measured/ changed 1

State at least one other variable to be controlled 1

Acknowledgements

Permission has been sought from all relevant copyright holders and Hodder Gibson is grateful for the use of the following:

Image © Edwin Verin/Shutterstock.com (Model Paper 1 Section 1 page 10);
Image © jupeart/Shutterstock.com (Model Paper 1 Section 2 page 25);
Image © Joggie Botma/Shutterstock.com (Model Paper 1 Section 2 page 27);
Image © Rich Carey/Shutterstock.com (Model Paper 2 Section 2 page 8);
Image © Alexander Gordeyev/Shutterstock.com (Model Paper 2 Section 2 page 23);
Image © LingHK/Shutterstock.com (Model Paper 2 Section 2 page 24);
Image © Morphart Creation/Shutterstock.com (Model Paper 3 Section 2 page 25);
Image © Stuart Elflett/Shutterstock.com (2014 Section 2 page 6);
Image © Ints Vikmanis/Shutterstock.com (2014 Section 2 page 21);
Image © Sandra R. Barba/Shutterstock.com (2014 Section 2 page 24);
Image © Rob Byron/Shutterstock.com (2015 Section 1 page 10);
Image © MarcelClemens/Shutterstock.com (2015 Section 2 page 23);
Image © Procy/Shutterstock.com (2015 Section 2 page 25).

Hodder Gibson would like to thank SQA for use of any past exam questions that may have been used in model papers, whether amended or in original form.